Dynamics Days Latin America and the Caribbean 2018

Dynamics Days Latin America and the Caribbean 2018

Special Issue Editors

Nicolás Rubido

Arturo C. Martí

MDPI • Basel • Beijing • Wuhan • Barcelona • Belgrade

MDPI

Special Issue Editors
Nicolás Rubido
Instituto de Física de Facultad de Ciencias (IFFC), Universidad de la República (UdelaR)
Uruguay

Arturo C. Martí
Instituto de Física de Facultad de Ciencias (IFFC), Universidad de la República (UdelaR)
Uruguay

Editorial Office
MDPI
St. Alban-Anlage 66
4052 Basel, Switzerland

This is a reprint of articles from the Special Issue published online in the open access journal *Mathematical and Computational Applications* (ISSN 2297-8747) in 2019 (available at: https://www.mdpi.com/journal/mca/special_issues/DDAYS_LAC_2018)

For citation purposes, cite each article independently as indicated on the article page online and as indicated below:

LastName, A.A.; LastName, B.B.; LastName, C.C. Article Title. *Journal Name* **Year**, *Article Number*, Page Range.

ISBN 978-3-03921-503-4 (Pbk)
ISBN 978-3-03921-504-1 (PDF)

Contents

About the Special Issue Editors vii

Preface to "Dynamics Days Latin America and the Caribbean 2018" ix

Santiago Boari, Gonzalo Uribarri, Ana Amador and Gabriel B. Mindlin
Observable for a Large System of Globally Coupled Excitable Units
Reprinted from: *Math. Comput. Appl.* **2019**, *24*, 37, doi:10.3390/mca24020037 1

Roberto C. Budzinski, Bruno Rafael Reichert Boaretto, Thiago L. Prado and Sergio Roberto Lopes
Investigation of Details in the Transition to Synchronization in Complex Networks by Using Recurrence Analysis
Reprinted from: *Math. Comput. Appl.* **2019**, *24*, 42, doi:10.3390/mca24020042 14

B. R. R. Boaretto, R. C. Budzinski, T. L. Prado and S. R. Lopes
Suppression of Phase Synchronization in Scale-Free Neural Networks Using External Pulsed Current Protocols
Reprinted from: *Math. Comput. Appl.* **2019**, *24*, 46, doi:10.3390/mca24020046 30

Rodrigo Simile Baroni, Ricardo Egydio de Carvalho, Bruno Castaldi and Bruno Furlanetto
Time Recurrence Analysis of a Near Singular Billiard
Reprinted from: *Math. Comput. Appl.* **2019**, *24*, 50, doi:10.3390/mca24020050 42

Estefanía Piegari and Silvina Ponce Dawson
Functional Ca^{2+} Channels between Channel Clusters are Necessary for the Propagation of IP_3R-Mediated Ca^{2+} Waves
Reprinted from: *Math. Comput. Appl.* **2019**, *24*, 61, doi:10.3390/mca24020061 58

Johan Llamoza and Desiderio A. Vasquez
Structures and Instabilities in Reaction Fronts Separating Fluids of Different Densities
Reprinted from: *Math. Comput. Appl.* **2019**, *24*, 51, doi:10.3390/mca24020051 72

Gilberto M. Nakamura, Brenno C. T. Cabella and Alexandre S. Martinez
Exact Evaluation of Statistical Moments in Superradiant Emission
Reprinted from: *Math. Comput. Appl.* **2019**, *24*, 66, doi:10.3390/mca24020066 88

Brenno Cabella, Fernando Meloni and Alexandre S. Martinez
Inadequate Sampling Rates Can Undermine the Reliability of Ecological Interaction Estimation
Reprinted from: *Math. Comput. Appl.* **2019**, *24*, 48, doi:10.3390/mca24020048 97

Gilberto M. Nakamura, Ana Carolina P. Monteiro, George C. Cardoso and Alexandre S. Martinez
Finite Symmetries in Agent-Based Epidemic Models
Reprinted from: *Math. Comput. Appl.* **2019**, *24*, 44, doi:10.3390/mca24020044 104

María Susana Torre, Jean-Marc Boyer and Jorge Tredicce
Influence of a Modulated Parameter on Hantavirus Infection
Reprinted from: *Math. Comput. Appl.* **2019**, *24*, 68, doi:10.3390/mca24030068 121

About the Special Issue Editors

Nicolás Rubido received his Ph.D. (2014) degree from the University of Aberdeen and his thesis was nominated by the University as an outstanding thesis and won the Springer Theses award. He received his B.Sc. (2008) and M.Sc. (2010) degrees from the Universidad de la República, where he has held a permanent Adjunct Professor (full-time) position at the Physics Institute since 2015. His research interests include nonlinear dynamics, complex systems and networks, and data analysis.

Arturo C. Martí is Professor of Physics at the Universidad de la República (Uruguay). He completed his Ph.D. in physics at the Universitat de Barcelona in 1997. His research interests are focused on traditional academic topics centered on fluids and nonlinear physics, but he has also recently become involved in science popularization programs, the organization of the Physics Olympiads, photo contests, and teacher training workshops. Recently, he has been developing physics experiments using smartphones and new technologies.

Preface to "Dynamics Days Latin America and the Caribbean 2018"

This book is devoted to present articles dealing with an exciting field of dynamical systems, nonlinear dynamics, and its wide spectrum of cross-disciplinary applications, such as in topics ranging from epidemiology or ecology to engineering and neural systems. These works were presented as communications at the conference "Dynamics Days Latin America and the Caribbean 2018" held in Punta del Este, Uruguay, from December 26th to 30th of 2018. The focus of the conference was the wide spectra of cross-disciplinary applications of nonlinear dynamics and the most useful and novel techniques for dealing with these problems. It was a great opportunity to gather a group of scientists within multidisciplinary fields with these common interests and objectives and to create the synergy necessary for effectively implementing the most diverse applications.

In the first article of this book (https://www.mdpi.com/2297-8747/24/2/37), Santiago Boari and his collaborators introduce a ground-breaking model (following the seminal work by Ott and Antonsen) to describe the global synaptic activation in large coupled excitable systems. The model reproduces realistic features of the synaptic coupling in real neural systems and, in particular, allows them to calculate macroscopic quantities (order parameters), which they show to be related to the local field potentials of experimentally recorded nervous systems.

In the second article, Roberto C. Budzinsk et al. (https://www.mdpi.com/2297-8747/24/2/42) uses recurrence analysis to study the transition to synchronization in complex networks of bursting oscillators. Focusing on small-world and scale-free networks, they report the emergence of non-stationary states and intermittency in the transition region of coupled Rulkov neuron maps; namely, as the coupling strength between the maps is increased and before the system achieves complete synchrony.

Thirdly, Bruno Rafael R. Boaretto et al. (https://www.mdpi.com/2297-8747/24/2/46) presents two methods for suppressing the anomalous phase synchronization emerging in scale-free neural network models. These methods are related to the deep-brain stimulation and delayed-feedback control and could potentially help controlling neurological disorders—such as epilepsy or Parkinson's neuropathies.

The article by Rodrigo Simile Barone et al. (https://www.mdpi.com/2297-8747/24/2/50) is devoted to the analysis of an eccentric annular billiard using time recurrence analysis. Billiards have been paradigmatic examples of rich dynamical behaviors—typically emerging in Hamiltonian systems—and with broad applications, such as in Plasma Physics.

Entering directly into the Biophysics of cell signaling, the fifth article by Estefanía Piegari and Silvina Ponce Dawson (https://www.mdpi.com/2297-8747/24/2/61) shows the specificity and universality of intracellular calcium signals by presenting a quantitative comparison between experiments and an innovative model based on excitable systems.

In the field of Fluid Dynamics, the contribution by Johan Llamoza and Desiderio A. Vasquez (https://www.mdpi.com/2297-8747/24/2/51) reviews interesting aspects of wave propagation in reaction-diffusion fronts, where they study the interaction between density-driven convection and fronts with diffusive instabilities. They report that the presence of density gradients in fronts governed by the Kuramoto–Sivashinsky equation could enhance or suppress complex behavior emerging in these reaction-diffusion systems.

The seventh article in the book, by Gilberto M. Nakumara et al. (https://www.mdpi.com/2297-8747/24/2/66), deals with Bose–Einstein condensates. In particular, with the superradiance, i.e., the coherent collective radiation caused by the interaction between many emitters that is mediated by a shared electromagnetic field. The novelty stems from a simplification of the model that explains superradiance, allowing the authors to calculate the statistical moments of the phenomenon.

Next, Brenno Cabel and collaborators (https://www.mdpi.com/2297-8747/24/2/48) show how important it is to adequately choose the sampling rates in population dynamics observations. This is particularly relevant in order to avoid undermining the reliable estimation of ecological interaction. In particular, they show that by choosing slow acquisition rates, the collected data can produce deceptive patterns, such as the prey becoming the predator.

An application of how stochastic processes, agent-based modeling, and symmetries can be used to the analysis of epidemic spreading is presented by Gilberto M. Nakamura et al. (https://www.mdpi.com/2297-8747/24/2/44) in the ninth article of this book. The authors present an algorithm that explores permutation symmetries to enhance the computational performance of agent-based epidemic models and show the statistical properties of propagation.

The book finishes with the work by María S. Torre et al. (https://www.mdpi.com/2297-8747/24/3/68), who propose an ecological model describing the temporal evolution of the infection of mice due to hantavirus and, therefore, with clear implications to public health.

We would like to acknowledge the help of Luisa Parodi and Sociedad Uruguaya de Física, who provide us with the photograph of the cover.

Nicolás Rubido, Arturo C. Martí
Special Issue Editors

Mathematical and Computational Applications

MDPI

Article

Observable for a Large System of Globally Coupled Excitable Units

Santiago Boari *,†, Gonzalo Uribarri *,†, Ana Amador and Gabriel B. Mindlin

Departamento de Física, Facultad de Ciencias Exactas y Naturales (FCEyN), UBA and IFIBA CONICET, Buenos Aires C1428BFA, Argentina; anita@df.uba.ar (A.A.); gabo@df.uba.ar (G.B.M.)

* Correspondence: santiagoboari@gmail.com (S.B.); gonzauri@gmail.com (G.U.)

† These authors contributed equally to this work.

Received: 27 February 2019; Accepted: 4 April 2019; Published: 6 April 2019

Abstract: The study of large arrays of coupled excitable systems has largely benefited from a technique proposed by Ott and Antonsen, which results in a low dimensional system of equations for the system's order parameter. In this work, we show how to explicitly introduce a variable describing the global synaptic activation of the network into these family of models. This global variable is built by adding realistic synaptic time traces. We propose that this variable can, under certain conditions, be a good proxy for the local field potential of the network. We report experimental, in vivo, electrophysiology data supporting this claim.

Keywords: local field potential; mean field models; coupled oscillators; theta neuron; synchrony; out of equilibrium system

1. Introduction

The behavior of large ensembles of out of equilibrium units is far from being completely understood. Recently, some bridges have been built to connect the dynamics of individual units with the collective state of a network (see, for example, [1–4]). This line of work has a long and rich history that includes the pioneering work of Art Winfree, who presented the first mathematical models built to describe the synchronization between biological oscillators [5,6]. Yoshiki Kuramoto also made important advances in this line of work. He proposed a simple model for the behavior of a large set of coupled oscillators, interacting pairwise through a sinusoidal function of their phase differences [7,8]. In this approach, the collective behavior of the system is described in terms of a single complex number: its amplitude accounts for the phase-coherence of the population of oscillators, and its phase stands for the average phase. In a typical statistical approach, the assumption is that this problem can be well approximated by a continuous system, described in terms of a density function of phase and time. This density represents the distribution of oscillators that, at a given time, present a given phase θ.

Ott and Antonsen proposed an approach to this problem that turned out to be a breakthrough in the field [9,10]. In that work, the authors studied the dynamics of a network of coupled oscillators. Each oscillator was described in terms of its phase. The continuity equation satisfied by the density describing the state of the network was decomposed in modes, and under a specific set of hypotheses, the amplitudes of the modes were found to be linked by a simple function. In this way, knowing the dynamics of the first mode was enough to reconstruct the behavior of the infinite set of modes. Moreover, with this approach, it is possible to show that an order parameter describing the synchronicity of the network might obey a low dimensional system of ordinary differential equations (the order dependent on the complexity of the network).

This reduction in the dimensionality of the system of equations for the mode amplitudes was only a good approach if the interaction between the units could be written in terms of specific functional forms, such us the Kuramoto coupling term (a sinusoidal function of the difference between the interacting phases). It also worked if the units presented excitable dynamics before coupling and if the coupling was modeled in terms of "pulse" functions [11]. Due to the similarity in their dynamics, this framework is generally used to model neural networks. The coupling describes the input current into the excitable units I as:

$$I = \frac{1}{N}\sum_{j}(1 - \cos(\theta_j)), \tag{1}$$

which will account for the contributions to the current I by the j units, as their phases pass the value $\theta_j \sim \pi$ (defined as the phase in which the neurons "spike"). Even if the dynamics of the individual units before the coupling was excitable, the couplings previously described are not the most natural ones to model synapses. In order to overcome this difficulty, Montbrió and collaborators derived independently exact equations to describe macroscopically networks of spiking units [12,13]. They were interested in the mechanisms of individual spike generation, and how they introduce an effective coupling between the mean membrane potential and the spiking rate, two biophysically relevant macroscopic quantities. In this approach, the firing rate is a good approximation of the global synaptic current for fast synapses. Moreover, to account for slower synapses, they proposed that the global synaptic activation S would be ruled by:

$$\tau\frac{dS}{dt} = -S + R, \tag{2}$$

where R stands for the spiking rate (the number of spikes occurring per unit of time) [12]. This approach led the authors to show that inhibitory, all-to-all coupled excitable units can display oscillations. This is a result that a Wilson-Cowan-like phenomenological model cannot reproduce [14].

In the first part of this work (Section 2), we build on these previous efforts by modeling the global synaptic activation as the sum of synaptic currents which present a maximum that is delayed with respect to the spike responsible for its occurrence. This is an important feature of the synaptic interaction that is not reproduced by previous models. Biophysical models of the synaptic interaction (nonlinear kinetic models [15]) have solutions that display this delay. Nevertheless, it is not known how to achieve an analytical macroscopic description of the system with these nonlinear equations describing the synaptic interactions. In this work, we present a model capable of reproducing this realistic feature of the synaptic coupling, but which is also compatible with the analytic calculation of macroscopic quantities for the network.

The ideas proposed by Ott and Antonsen were successfully applied to study the $N \to \infty$ limit in different kinds of networks of coupled units [16–21]. In cases where the composing elements of the system are excitable, such as neurons, an order parameter describing the synchrony of the network can indicate a highly synchronous state either because the units are spiking in phase, or because the units are quiescent near each other [22]. To account not only for its synchrony but also for its level of activity, different quantities have been proposed to describe the global state of a network. Yet, these quantities are not unrelated. In recent work, it has been shown that the spiking rate of the network can be analytically expressed as a function of its synchrony [11]. A different approach was followed by Montbrió et al., who formulated a model for a network of quadratic integrate-and-fire units (QIF) in terms of its average voltage and its firing rate [23]. Both approaches (a network of phase oscillators and a network of QIF neurons) have been proven to be equivalent. Yet, despite the clear interpretation of the firing rate as a variable describing the state of a network, its direct measurement constitutes a challenge, as it requires recording the individual activity of every neuron in the network. In the second part of this work (Sections 3 and 4), we discuss how the macroscopic variable defined in Section 2 compares to the local field potential (LFP), and we test this relationship using measurements in an actual nervous system.

2. Macroscopic Evolution of a Set of Excitable Units

Let us assume a network of N excitable units whose dynamics are described in terms of the phases θ_i, $i = 1 \dots N$, obeying:

$$\frac{d\theta_i}{dt} = (1 - \cos(\theta_i)) + (1 + \cos(\theta_i))(\eta_i + J\,S), \tag{3}$$

where η_i defines the degree of excitability of the ith unit, J the coupling strength between the units, and S stands for the average synaptic current between the units. This model is known as the theta neuron model, and it is a simple one-dimensional model for the spiking of a neuron [24]. The variable θ lies on the unit circle and ranges between 0 and 2π. When $\theta = \pi$ the neuron "spikes", that is, it produces an action potential.

As we show in Appendix A, when taking the continuous limit, the order parameter $z = \sum_i e^{i\theta_i}$, obeys the following dynamical rule:

$$\frac{dz}{dt} = iz(1 + \eta_0 + JS) - z\Delta + \frac{i}{2}(1 + \eta_0 + JS)\left(1 + z^2\right) - \frac{\Delta}{2}\left(1 + z^2\right) \tag{4}$$

with $S = \sum_i s_i$, where each s_i describes the synaptic current contributed by a neuron spiking at $t = t_i$ and η_0 and Δ are the mean and width of a Lorentzian distribution for $g(\eta)$, the excitability distribution function of the population (see details in Appendix A). If we assume that each synaptic current can be represented by a function:

$$s_i \propto \begin{cases} (t - t_i)e^{-\frac{t - t_i}{\tau}}, & \text{if } t > t_i \\ 0, & \text{if } t \leq t_i \end{cases} \tag{5}$$

we can write a two-dimensional linear dynamical system having this function as a solution [25], which reads:

$$\tau \frac{ds_i}{dt} = -s_i + x_i \tag{6}$$

$$\tau \frac{dx_i}{dt} = -x_i + \delta(t - t_i). \tag{7}$$

Since these equations are linear, the global variable S will satisfy the same equations, with the activity $\phi(t)$ (the total number of spikes taking place per unit of time) as the forcing term. This can be expressed in terms of the order parameter as (see detailed calculation in Appendix A):

$$\phi(t) = \frac{2}{\pi}\left(\frac{1 + Re\, z}{|1 + z|^2} - \frac{1}{2}\right). \tag{8}$$

In this way, the network of coupled units can be macroscopically described by the following system:

$$\frac{dz}{dt} = iz(1 + \eta_0 + JS) - z\Delta + \frac{i}{2}(1 + \eta_0 + JS)\left(1 + z^2\right) - \frac{\Delta}{2}\left(1 + z^2\right) \tag{9}$$

$$\tau \frac{dS}{dt} = -S + x \tag{10}$$

$$\tau \frac{dx}{dt} = -x + \frac{2}{\pi}\left(\frac{1 + Re\, z}{|1 + z|^2} - \frac{1}{2}\right). \tag{11}$$

In this calculation, we have assumed a unique population of neurons (for example all excitatory ones), with parameters distributed in a Lorentzian way (mean excitability parameter η_0), and with all the units coupled among each other with a unique strength described by J. For this simple architecture, we illustrate the different solutions of the problem in Figure 1. The calculations used to write these equations are presented in Appendix A.

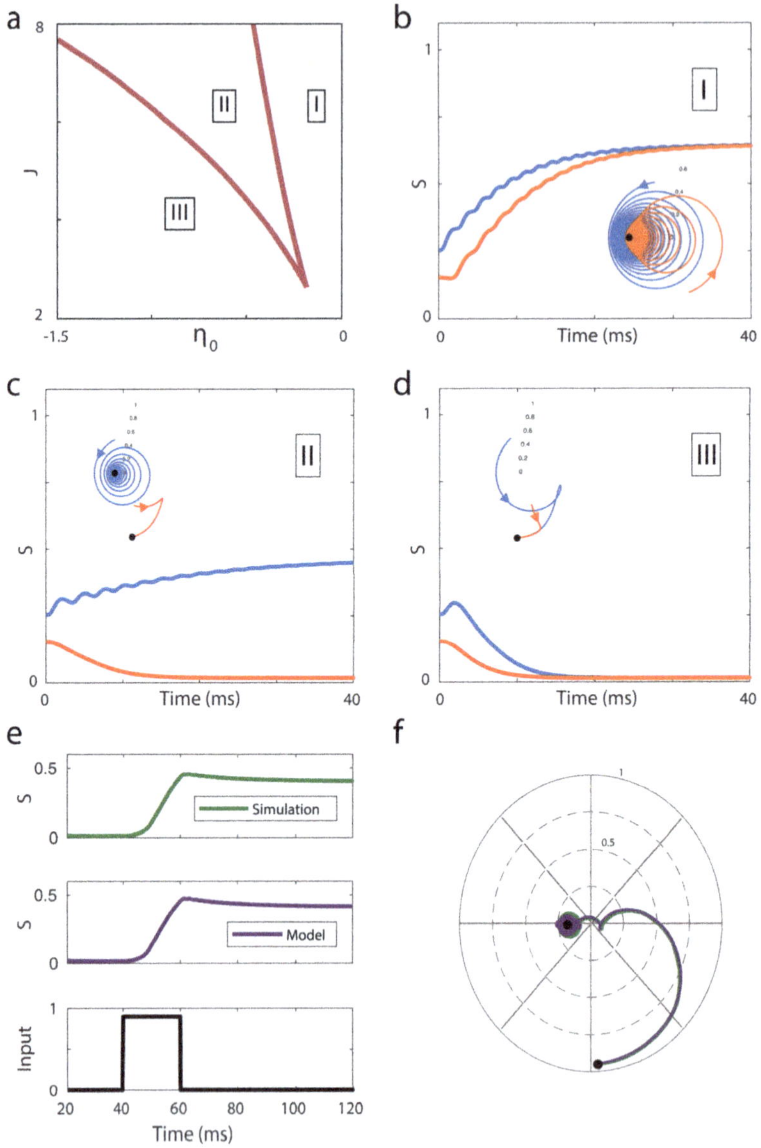

Figure 1. Solutions of the proposed model for the case of a unique excitatory population. (**a**) Bifurcation diagram of the system in the (η_0, J) plane for a case with $\tau = 2$ ms and $\Delta = 0.1$. The curves denote saddle node bifurcations. Three regions corresponding to the different types of solutions can be identified. (**b–d**) Simulations for different initial conditions. Regions I and III present a unique stationary attracting solution, while region II of parameter space presents bi-stability. Insets show the evolution of the order parameter of the system, with black dots representing the attractive fixed points. (**e**) Agreement between the reduced model and a simulation of a system of oscillators (15,000 units, using a time step of $\Delta t = 0.001$ s)). The rate used to drive the system was computed as the number of oscillators crossing the value $\theta = \pi$, divided by the total number of oscillators and the time step Δt. (**f**) Evolution of the order parameter for the model and the simulation.

The panel (a) in Figure 1 displays three different regions of the parameter space (η_0, J), for a problem in which $\tau = 2$ ms and $\Delta = 0.1$. The three panels displayed in b–d correspond to simulations for different initial conditions, showing that regions I and III present a unique stationary attracting solution. Region II presents bi-stability. This result is consistent with what is expected from simple additive models [26]. Panels (e) and (f) in Figure 1, show the agreement between this macroscopic description of the system, and a simulation of 15,000 oscillators, whose dynamics are ruled by our original set of equations. The rate used to drive the system was computed as the number of oscillators crossing the value $\theta = \pi$, divided by the total number of oscillators and the time step Δt.

Notice that in our description, the variable z describes the system´s synchronicity, and the variable S represents the global synaptic activation of the network. The individual synaptic currents are in fact the result of nonlinear processes. But since the functional form $s_i \sim (t - t_i)e^{-\frac{t-t_i}{\tau}}$ is a most successful fit for a synaptic current [15], we use a linear model for its generation. This allows us to add the equations for each synaptic component in order to come up with an equation of the global synaptic activation. Remarkably, synaptic activity is often acknowledged as the most important source of extracellular current flow [27]. In the next sections, we will show that it is possible to use electrophysiological measurements as a starting point to compute a variable that can be a proxy for the synaptic activation.

3. An Experimental System

Synaptic currents are conjectured to contribute in a substantial way to the LFP since extracellular currents from many individual compartments must overlap in time to induce a measurable signal. This requires pertinent events to be slow [27]. In general, complex neural architectures involve both excitatory and inhibitory neurons. This poses a problem for reconstructing the origin of any given fluctuation in the LFP. However, synchronous action potentials from many neurons can contribute substantially at specific temporal instances, particularly in the cases where the structure of the network allows us to have inhibitory and excitatory neurons spiking out of phase. Songbirds have been shown to present this out-of-phase spiking between excitatory and inhibitory neurons [28]. We will thus investigate a system with complex neural architecture but for which, during some time intervals, mostly one class of neurons are active. We will then concentrate on those time intervals and investigate whether a reconstructed global synaptic activity can approximate the recorded LFP.

Songbirds have highly specialized brain regions to generate and process the signals that are involved in song production and perception. In a specific region of the telencephalon (known as the nucleus HVC, an old acronism at present used as a proper name), some neurons are active during song production. Interestingly, those neurons also spike when the bird is asleep or anesthetized if it is exposed to a recording of its own song (e.g., [29]). Moreover, a neuron that spikes at a specific temporal instance when producing song will spike at about the same temporal instance when the anesthetized or sleeping bird listens to the song recording [30]. This paradigm motivates the study of auditory-elicited responses in the HVC and its link to the coding of vocal production.

For the data presented in this work, extracellular recordings of neural activity were conducted on urethane-anesthetized canaries (*Serinus canaria*). Surgery was performed to access the neural nuclei HVC and insert a multi-electrode array (A1 $\times$ 32, Neuronexus Technologies, Inc.). This array contained 32 aligned electrodes, separated 25 μm from each other. Neural activity was monitored online using proprietary INTAN software to control an Intan RHD2000 acquisition board. To study auditory responses in HVC, the experimental protocol consisted of presenting three different auditory stimuli (BOS, bird's own song; CON, the song of an adult male conspecific; REV, its own song in reverse). Each protocol consisted of twenty randomized presentations of each stimulus (for more detailed methods, see [31]). These methods are the standard paradigm to study selectivity in the neuronal nucleus HVC. Signals were sampled at 20 kHz and the hardware filtering was set between 0.1 Hz and 5000 Hz.

Recordings were analyzed using custom-built software. Low-frequency components due to synaptic currents were isolated from the high-frequency spiking activity due to action potentials elicited by neurons near each recording site. The slow signals commonly referred to as LFP were obtained using a low-pass zero-phase Butterworth IIR digital filter on the raw data (4th order, cutoff frequency: 300 Hz). For spiking activity, the filter used was a high-pass zero-phase Butterworth IIR digital filter (4th order, cutoff frequency: 500 Hz).

Figure 2 shows examples of the high-pass filtered data from one protocol. These traces represent the neural response to auditory presentations of the bird's own song. In Figure 2a, we show the sound signal from a single canary syllable (part of the auditory stimulus that was presented to the anesthetized bird). In Figure 2b, we show a segment from 10 traces of high-pass filtered data. These traces correspond to the recorded activity in one channel of the neural probe for 10 presentations of the bird's own song. In Figure 2c, we show a magnified example of presentation 1 in Figure 2b and the threshold used for spike detection. As can be seen from the different traces in Figure 2b,c, there are multiple spikes of different amplitudes. This is an indicator that the electrode is registering multiunit activity (i.e., spikes from different neurons located at different positions from the electrode). As we are registering extracellular spikes, the amplitude registered by the electrode decays with distance. For additional details on the recording of neural ensembles, see [32]. A simple characterization of the overall neural response to the stimulus is given by the multiunit activity histogram, which is computed by thresholding the signal, detecting the times at which each spike occurred and binning the activity in 15 ms windows.

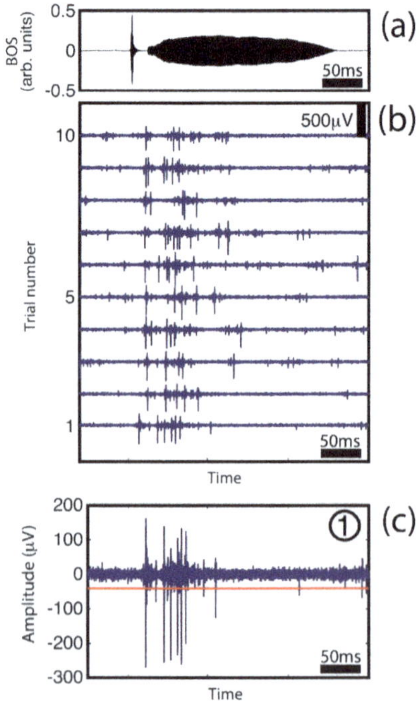

Figure 2. Raw data thresholding and multiunit activity (MUA). (**a**) BOS sound segment (single canary song syllable). (**b**) High-pass filtered raw data traces of the neural response to auditory presentations of BOS in anesthetized birds (see Section 3). Traces correspond to 10 trials from one protocol. Each trace consists of background electrical noise and sharp spikes corresponding to the extracellular recording of an action potential. The threshold allows the detection of spikes of multiple amplitudes. Differences in spike shape and amplitudes correspond to the electric activity generated by different neurons. Thus, the activity obtained by thresholding is multiunit in nature. After thresholding, the spikes are treated as a series of timestamps of where spiking occurred. (**c**) Zoomed-in trace for trial 1, showing the threshold level for detection.

Figure 3 illustrates the results from the protocol from which the segment shown in Figure 2 was extracted. The top panel in Figure 3a shows the BOS recording presented to the anesthetized bird. The average LFP trace (trial-averaged for 20 trials and channel-averaged for the 32 channels) is shown in the second panel. This average was computed to consider all the synaptic currents in the recording, which is required for comparison with the global synaptic activation S that we will reconstruct from these data. Since this signal represents the average, note that peaks arise both from the robustness in the response (trial-average) and from the synchronization of multiple channels (channel-average).

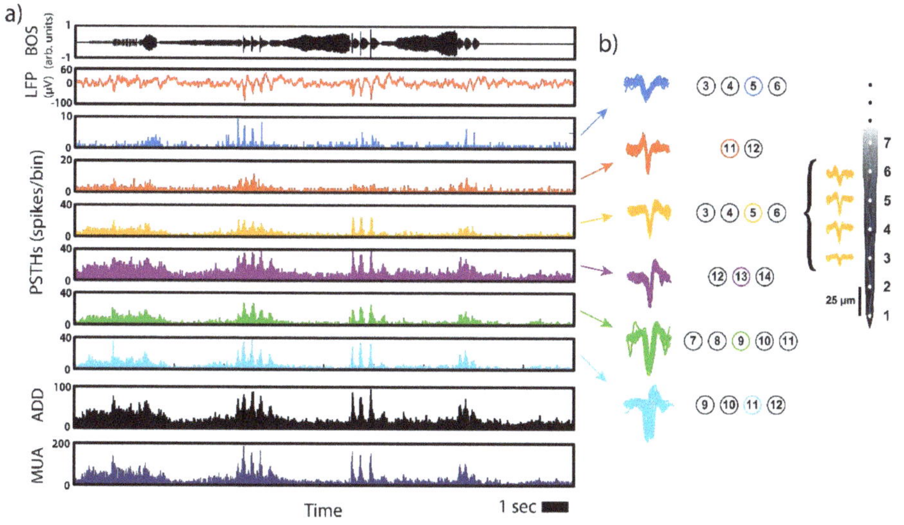

Figure 3. Single unit activity is synchronized in multiple recording sites. (**a**) Activity profiles of spike-sorted clusters from a recording. From top to bottom: BOS sound signal, trial- and channel-averaged LFP, PSTHs for each neuron and summed single unit activity (ADD). PSTHs are histograms (15 ms bins) of the activity elicited in each isolated neuron during 20 auditory presentations of the BOS. Lastly, the bottom panel shows the multiunit activity profile (MUA), obtained by thresholding the recorded neural data (see Section 3 and Figure 2). (**b**) Each action potential is recorded by several channels in the multielectrode array (a diagram is shown on the right). Spikes from individual, well-isolated neurons are shown with different colors. Spikes from the same neuron are simultaneously recorded as spikes of different shapes in different channels (see yellow cluster inset). The channels where each cluster was detected are shown to the right of each spike group. Color outlines indicate the maximum amplitude channel for each cluster, which corresponds with the spike shapes shown. Each cluster presents a robust response across trials (sharp peaks present in the PSTHs in (**a**)). Additionally, these results show that registered neurons are synchronized. The sharp peaks in the ADD profile in (**a**) result from the combination of response robustness across trials and from the synchronous firing of isolated neurons.

Single neuron activity was also recovered from the recordings. A spike-sorting algorithm (Phy, [33]) identifies the temporal instances where different neurons are spiking by conducting a principal component analysis (PCA) on detected spike shapes. For the data shown in Figure 3, identified neurons are plotted with different colors (see Figure 3b). The same neuron could be registered simultaneously in more than one channel (as an example, we show in Figure 3b the yellow spike shape registered in several channels of the multi-electrode). The circles to the right of each set of spike shapes indicate the channels where that neuron was registered. We have also color-outlined the circles representing the maximum amplitude channel, which correspond to the spike shapes shown. Binning the spike times for each neuron with 15 ms bins, we get the time traces displayed in the insets

3–8 in Figure 3a (post-stimulus time histograms or PSTH). Peaks within each of these signals account for the response robustness at a given temporal instance and the presence of higher activation levels in response to the song. Notice that, additionally, whenever these time traces show peaks, their positions coincide, meaning that there is a degree of synchronization among units in the part of the neural network that is being sampled in these recordings.

Only neurons that respond mostly to the bird´s own song (BOS) were taken into account in our analysis. Some neurons present brief bursts of activity at a few instances during the production of a specific syllable type. Other neurons spike in a tonic-like fashion. Comparison between these firing patterns [34], and results from another species (zebra finches, *Taeniopygia guttata* [35]), suggests that the first type might be projection neurons, while the second class might correspond to inhibitory interneurons. Projection neurons are excitatory neurons [36]. Neurons shown in Figure 3 are putative projection neurons, since they present bursts of activity at specific instances within the song.

4. Reconstructing the Dynamics for S from the Data

The single-channel multiunit activity recorded when the anesthetized bird is exposed to its own song, summed over the different repetitions, presents clear peaks at specific temporal instances (see Figure 3). We have also found that at least close to the peaks, this MUA is similar to the summed activity of several different neurons detected by the multi-electrode at different depths of the nucleus HVC (compare MUA and ADD in Figure 3a). All these neurons spiking at the same temporal instances are of the same kind (either excitatory or inhibitory), and therefore will contribute additively to the average synaptic current. In this way, we can define a threshold for the multiunit activity, and identify the times t_i where spikes are detected (as was shown in Figure 2c). With that sequence of times t_i, $i = 1 \ldots N$, we can add the functions for each synaptic event s_i:

$$s_i \propto \begin{cases} (t - t_i)e^{-\frac{t-t_i}{\tau}}, & \text{if } t > t_i \\ 0, & \text{if } t \leq t_i, \end{cases} \tag{12}$$

and build a proxy for the average synaptic current, at least close to the instances where the multiunit activity has peaks. Using a parsimonious estimation for excitatory synapses of $\tau = 10$ ms, [15] we generate this synthetic activation S, and compare the signal with the local field potential. The result is shown in Figure 4. In blue (top panel), the trace shows S as reconstructed from the spiking times and the red trace (bottom panel) is the LFP obtained from the recordings (also shown in Figure 3a). These two traces share some temporal features: the instances where S presents peaks correspond to the peaks found in the LFP. However, the LFP presents additional variations that S does not capture. Most probably, this is because S is reconstructed with a binned version of the high-passed data, where only the instances of the supra-threshold spiking activity were considered. This, in turn, yields a time trace for S that presents small fluctuations, while the measured LFP presents additional variations arising from the background electrical activity. Finally, to measure the similarity between the two, we computed the average Pearson correlation between the two signals in shifting windows of 1.0 s, and we got $c_{\text{mean}} \sim 0.47$, with correlation values reaching $c_{\text{max}} \sim 0.86$ in the regions with the peaks. This informs that the reconstructed S is more reliable in the case where synchronous firing has occurred, such that an emerging pattern can be observed from the data.

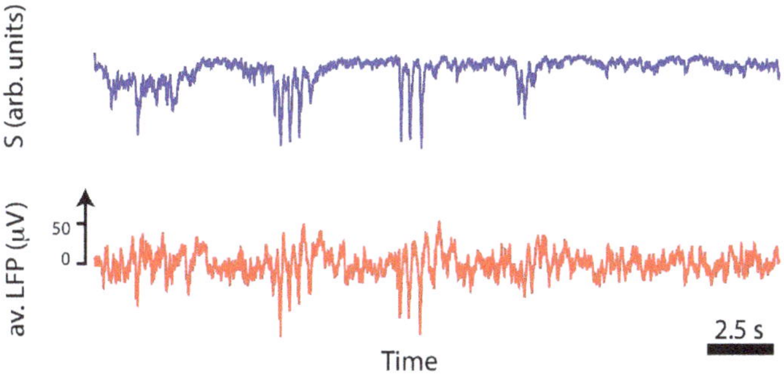

Figure 4. The reconstructed global synaptic activation captures prominent LFP features. Time traces for the global synaptic activation S (top panel, blue) and the trial- and channel- averaged LFP (bottom panel, red). The trace obtained for S approximates the measured LFP, especially where large peaks occur. The similarity between the two signals was measured using Pearson's correlation coefficient, which yields a maximum value of $c_{max} \sim 0.86$ by taking the regions with the peaks and $c_{mean} \sim 0.47$ for the whole timespan. The difference in correlation strength means that the reconstructed S better approximates the LFP near signal events that correspond to the synchronous firing of multiple neurons (see Figure 3).

5. Discussion

In recent years, it has been shown that, in the infinite size limit, certain systems of globally coupled phase oscillators can display macroscopic features that obey low dimensional dynamics. That class of systems includes excitable systems, and therefore it is natural to inquire about its consequences in neuroscience, studying how large sets of neurons synchronize to generate behavior. Different functional forms describing the interaction between the units were studied in order to achieve an analytic macroscopic description of a network. In this work, we built a model for the global synaptic activation of a neural network, by adding functions that represented realistic synaptic currents. In particular, each synaptic current had a maximum that was delayed with respect to the maximum of the spike responsible for its occurrence. This led us to propose a two-dimensional linear model for each current, and therefore a two-dimensional model for the global activation, with the firing rate as its driving force. We computed and integrated the differential equations satisfied by quantities that describe the network macroscopically. Then, we simulated the networks directly and computed the same observables, finding a remarkable agreement.

It has been pointed out that in physiological situations, synaptic activity is often the most important source of extracellular current flow. This is because many events need to contribute to induce a measurable signal, which privileges slow events as synaptic currents. The effect is amplified if large synchrony exists. We tested the hypothesis that a global synaptic activation, reconstructed from the spikes detected by a multi-electrode assuming an excitatory nature, could approximate the LFP at some temporal instances. We did it for a system which presents an architecture far more complex than the one used to introduce our model. Nevertheless, we restrained our analysis to temporal intervals where large synchronic events of neurons of a single type are expected and found that, for those time intervals, the LFP data and the computed synaptic activation were highly correlated.

It is possible to obtain a significant amount of information from a dynamical system by measuring some or even one of its variables. For example, it has been shown that it is possible to reconstruct the topology of a dynamical system that displays chaotic behavior by building an embedding from one of the system´s variables [37]. Moreover, for some systems, it is possible to reconstruct their ruling equations by operating on one measured variable [38]. In this way, the similarity between LFP (measurable) and the synaptic global activation (used in our macroscopic models) suggests a path to build bridges between macroscopic models for large sets of excitable units and experimental data.

Author Contributions: Conceptualization G.B.M. and A.A.; methodology, software, validation and formal analysis, G.U., S.B., A.A. and G.B.M.; writing—original draft preparation, G.B.M. and A.A.; visualization, G.U. and S.B.

Funding: This work describes research partially funded by National Council of Scientific and Technical Research (CONICET, Argentina), National Agency of Science and Technology (ANPCyT, Argentina), University of Buenos Aires (UBA, Argentina) and National Institute of Health through R01-DC-012859.

Acknowledgments: Experiments were performed in accordance with a protocol approved by the University of Buenos Aires (FCEN-UBA) Institutional Animal Care and Use Committee.

Conflicts of Interest: The authors declare no conflict of interest.

Appendix A

In this Appendix, we will first derive the expression of the spiking rate of the network as a function of its order parameter. Secondly, we will derive the equation ruling the dynamics of the order parameter.

Let us assume a very large network of units described in terms of their phases θ_i, $i = 1 \ldots N$, obeying the following dynamical system:

$$\frac{d\theta_i}{dt} = (1 - \cos(\theta_i)) + (1 + \cos(\theta_i))(\eta_i + I(\{\theta_j\})), \tag{A1}$$

where $I(\{\theta_j\})$ represents the coupling function between the units. Let us assume that this term can be written in terms of the order parameter z of the population. Assuming an infinitely large number of oscillators, we propose a continuous description of the population, described in terms of a probability density $f(\theta, \eta, t)$ of the oscillators with parameter η being phase θ at time t. The spiking rate of the network can be computed as:

$$\phi(t) = \int_{-\infty}^{\infty} f(\theta, \eta, t) \frac{d\theta}{dt}(\theta, \eta, t)|_{\theta=\pi}\, d\eta. \tag{A2}$$

Notice that out of the two terms in the integral; the second one gets a very simple form:

$$\frac{d\theta}{dt}|_{\theta=\pi} = 2 + 0(\eta_i + I(z)) = 2. \tag{A3}$$

Concerning the probability density, we can write it as in [9]:

$$f(\theta, \eta, t) = \frac{g(\eta)}{2\pi}\left[1 + \sum_{n=1}^{\infty} \alpha(\eta, t)^n e^{in\theta} + \alpha^*(\eta, t)^n e^{-in\theta}\right]. \tag{A4}$$

In this way, since $e^{in\pi} = e^{-in\pi} = (-1)^n$, we can write:

$$\sum_{n=1}^{\infty} \alpha(\eta, t)^n (-1)^n = \frac{1}{1+\alpha} - 1, \tag{A5}$$

and

$$\sum_{n=1}^{\infty} \alpha^*(\eta, t)^n (-1)^n = \frac{1}{1+\alpha^*} - 1, \tag{A6}$$

which leads to:

$$f(\pi, \eta, t) = \frac{g(\eta)}{2\pi}\left[\frac{1}{1+\alpha} + \frac{1}{1+\alpha^*} - 1\right], \tag{A7}$$

and therefore

$$\phi(t) = \int_{-\infty}^{\infty} 2\frac{g(\eta)}{2\pi}\left[\frac{1}{1+\alpha} + \frac{1}{1+\alpha^*} - 1\right]d\eta \equiv \phi_1(t) + \phi_2(t) + \phi_3(t) \tag{A8}$$

The three terms, assuming a Lorentzian distribution for $g(\eta)$ with maximum at η_0 and width Δ, give (by using the Residue theorem to evaluate the integrals):

$$\phi_1(t) = \int_{-\infty}^{\infty} \frac{g(\eta)}{(1+\alpha(\eta,t))}d\eta = \frac{1}{\pi}\left(\frac{1}{1+\alpha(\eta_0 - i\Delta, t)}\right) \tag{A9}$$

$$\phi_2(t) = \int_{-\infty}^{\infty} \frac{g(\eta)}{(1+\alpha^*(\eta,t))}d\eta = \frac{1}{\pi}\left(\frac{1}{1+\alpha^*(\eta_0 + i\Delta, t)}\right) \tag{A10}$$

$$\phi_3(t) = -\int_{-\infty}^{\infty} \frac{g(\eta)}{\pi}d\eta = -\frac{1}{\pi}. \tag{A11}$$

On the other hand, the definition of the order parameter is:

$$z(t) = \int_{-\infty}^{\infty} f(\theta,\eta,t)e^{i\theta}d\eta = \int_{-\infty}^{\infty} \frac{g(\eta)}{2\pi}\alpha^*(\eta,t)d\eta = \alpha^*(\eta_0 + i\Delta, t), \tag{A12}$$

and therefore

$$\phi(t) = \frac{1}{\pi}\left(\frac{1}{1+z(t)} + \frac{1}{1+z^*} - 1\right) = \frac{2}{\pi}\left(\frac{1+Re\, z}{|1+z|^2} - \frac{1}{2}\right). \tag{A13}$$

In order to derive the equation of the order parameter, we start with the continuity equation satisfies by the probability density f:

$$\frac{\partial f}{\partial t} + \frac{\partial}{\partial \theta}\left(\dot{\theta} f\right) = 0. \tag{A14}$$

We can expand the probability density in terms of the angular modes:

$$f_n(\eta,\theta,t) = \frac{g(\eta)}{2\pi}\left(1 + \alpha^n e^{in\theta} + \alpha^{*n} e^{-in\theta}\right), \tag{A15}$$

where the distribution:

$$\Gamma(\eta) = \frac{\Delta/\pi}{\left((\eta-\eta_0)^2 + \Delta^2\right)} \tag{A16}$$

describes the distribution of the units' parameters. Replacing the expansion in the continuity equation, we get the following equation for the mode amplitudes:

$$\dot{\alpha} = -i\left[\alpha(1+\eta+JS) + \left(\alpha^2+1\right)\left(\frac{\eta+JS-1}{2}\right)\right]. \tag{A17}$$

Finally, this mode can be related to the order parameter. The reason is that

$$z = \int_{-\infty}^{\infty} f(\eta',\theta',t)e^{i\theta'}d\theta' d\eta' = \int_{-\infty}^{\infty} \alpha^*(\eta',t)\Gamma(\eta')d\eta' \tag{A18}$$

and therefore, using the theorem of the residues,

$$\dot{z} = i\,\alpha(1+\eta_0+JS) - z\Delta + i\left(1+z^2\right)\left(\frac{\eta_0+JS-1}{2}\right) - \frac{\Delta}{2}\left(1+z^2\right). \tag{A19}$$

References

1. Strogatz, S.H. From Kuramoto to Crawford: Exploring the onset of synchronization in populations of coupled oscillators. *Physica D* **2000**, *143*, 1–20. [CrossRef]
2. Dörfler, F.; Bullo, F. Synchronization in complex networks of phase oscillators: A survey. *Automatica* **2014**, *50*, 1539–1564. [CrossRef]
3. Chandra, S.; Hathcock, D.; Crain, K.; Antonsen, T.M.; Girvan, M.; Ott, E. Modeling the network dynamics of pulse-coupled neurons. *Chaos Interdiscip. J. Nonlinear Sci.* **2017**, *27*, 033102. [CrossRef]
4. Skardal, P.S.; Restrepo, J.G.; Ott, E. Uncovering low dimensional macroscopic chaotic dynamics of large finite size complex systems. *Chaos Interdiscip. J. Nonlinear Sci.* **2017**, *27*, 083121. [CrossRef] [PubMed]
5. Winfree, A.T. Biological rhythms and the behavior of populations of coupled oscillators. *J. Theor. Biol.* **1967**, *16*, 15–42. [CrossRef]
6. Winfree, A.T. *The Geometry of Biological Time*, 12th ed.; Springer: New York, NY, USA, 2001.
7. Acebrón, J.A.; Bonilla, L.L.; Vicente, C.J.P.; Ritort, F.; Spigler, R. The Kuramoto model: A simple paradigm for synchronization phenomena. *Rev. Mod. Phys.* **2005**, *77*, 137–185. [CrossRef]
8. Kuramoto, Y. Self-entrainment of a population of coupled non-linear oscillators. In *International Symposium on Mathematical Problems in Theoretical Physics*; Lecture Notes in Physics; Araki, H., Ed.; Springer: Berlin/Heidelberg, Germany, 1975; Volume 39, pp. 420–422.
9. Ott, E.; Antonsen, T.M. Low dimensional behavior of large systems of globally coupled oscillators. *Chaos* **2008**, *18*, 037113. [CrossRef] [PubMed]
10. Ott, E.; Antonsen, T.M. Long time evolution of phase oscillator systems. *Chaos* **2009**, *19*, 023117. [CrossRef]
11. Roulet, J.; Mindlin, G.B. Average activity of excitatory and inhibitory neural populations. *Chaos* **2016**, *26*, 093104. [CrossRef]
12. Devalle, F.; Roxin, A.; Montbrió, E. Firing rate equations require a spike synchrony mechanism to correctly describe fast oscillations in inhibitory networks. *PLoS Comp. Biol.* **2017**, *13*, e1005881. [CrossRef]
13. Schmidt, H.; Avitabile, D.; Montbrió, E.; Roxin, A. Network mechanisms underlying the role of oscillations in cognitive tasks. *PLoS Comp. Biol.* **2018**, *14*, e1006430. [CrossRef]
14. Wilson, H.R.; Cowan, J.D. Excitatory and inhibitory interactions in localized populations of model neurons. *Biophys. J.* **1972**, *12*, 1–24. [CrossRef]
15. Destexhe, A.; Mainen, Z.F.; Sejnowski, T.J. Synaptic currents, neuromodulation, and kinetic models. In *The Handbook of Brain Theory and Neural Networks*; Arbib, M.A., Ed.; MIT Press: Cambridge, MA, USA, 1995; pp. 956–959.
16. Childs, L.M.; Strogatz, S.H. Stability diagram for the forced Kuramoto model. *Chaos* **2008**, *18*, 043128. [CrossRef]
17. Restrepo, J.G.; Ott, E. Mean-field theory of assortative networks of phase oscillators. *Europhys. Lett.* **2014**, *107*, 60006. [CrossRef]
18. Laing, C.R. Exact neural fields incorporating gap junctions. *SIAM J. Appl. Dyn. Syst.* **2015**, *14*, 1899–1929. [CrossRef]
19. Rodrigues, F.A.; Peron, T.K.D.; Ji, P.; Kurths, J. The Kuramoto model in complex networks. *Phys. Rep.* **2016**, *610*, 1–98. [CrossRef]
20. Laing, C.R. Phase oscillator network models of brain dynamics. In *Computational Models of Brain and Behavior*; John Wiley & Sons, Ltd.: Hoboken, NJ, USA, 2017; pp. 505–517.
21. Uribarri, G.; Mindlin, G.B. Resonant features in a forced population of excitatory neurons. *arXiv*, **2019**, arXiv:1902.06008.
22. Luke, T.B.; Barreto, E.; So, P. Complete classification of the macroscopic behavior of a heterogeneous network of theta neurons. *Neural Comput.* **2013**, *25*, 3207–3234. [CrossRef] [PubMed]
23. Montbrió, E.; Pazó, D.; Roxin, A. Macroscopic description for networks of spiking neurons. *Phys. Rev. X* **2015**, *5*, 021028. [CrossRef]

24. Bard, E.G.; Kopell, N. Parabolic bursting in an excitable system coupled with a slow oscillation. *SIAM J. Appl. Math.* **1986**, *46*, 233–253.
25. Wilson, H.R. *Spikes, Decisions, and Actions: The Dynamical Foundations of Neurosciences*; Oxford University Press: New York, NY, USA, 1999.
26. Hoppensteadt, F.C.; Izhikevich, E.M. *Weakly Connected Neural Networks*; Springer: New York, NY, USA, 2012.
27. Buzsáki, G.; Anastassiou, C.A.; Koch, C. The origin of extracellular fields and currents—EEG, ECoG, LFP and spikes. *Nat. Rev. Neurosci.* **2012**, *13*, 407–420. [CrossRef] [PubMed]
28. Markowitz, J.E.; Liberti, W.A., III; Guitchounts, G.; Velho, T.; Lois, C.; Gardner, T.J. Mesoscopic patterns of neural activity support songbird cortical sequences. *PLoS Biol.* **2015**, *13*, e1002158. [CrossRef] [PubMed]
29. Margoliash, D. Preference for autogenous song by auditory neurons in a song system nucleus of the white-crowned sparrow. *J. Neurosci.* **1986**, *6*, 1643–1661. [CrossRef] [PubMed]
30. Dave, A.S.; Margoliash, D. Song replay during sleep and computational rules for sensorimotor vocal learning. *Science* **2000**, *290*, 812–816. [CrossRef] [PubMed]
31. Boari, S.; Amador, A. Neural coding of sound envelope structure in songbirds. *J. Comp. Physiol. A* **2018**, *204*, 285–294. [CrossRef]
32. Buzsáki, G. Large-scale recording of neuronal ensembles. *Nat. Neurosci.* **2004**, *7*, 446–451. [CrossRef]
33. Rossant, C.; Kadir, S.N.; Goodman, D.F.; Schulman, J.; Hunter, M.L.; Saleem, A.B.; Grosmark, A.; Belluscio, M.; Denfield, G.H.; Ecker, A.S. Spike sorting for large, dense electrode arrays. *Nat. Neurosci.* **2016**, *19*, 634–641. [CrossRef] [PubMed]
34. Del Negro, C.; Lehongre, K.; Edeline, J.M. Selectivity of canary HVC neurons for the bird's own song: Modulation by photoperiodic conditions. *J. Neurosci.* **2005**, *25*, 4952–4963. [CrossRef]
35. Hahnloser, R.H.; Kozhevnikov, A.A.; Fee, M.S. An ultra-sparse code underliesthe generation of neural sequences in a songbird. *Nature* **2002**, *419*, 65–70. [CrossRef]
36. Mooney, R. Different subthreshold mechanisms underlie song selectivity in identified HVc neurons of the zebra finch. *J. Neurosci.* **2000**, *20*, 5420–5436. [CrossRef]
37. Mindlin, G.M.; Gilmore, R. Topological analysis and synthesis of chaotic time series. *Physica D* **1992**, *58*, 229–242. [CrossRef]
38. Gouesbet, G. Reconstruction of vector fields: The case of the Lorenz system. *Phys. Rev. A* **1992**, *46*, 1784–1796. [CrossRef]

Mathematical and Computational Applications

MDPI

Article

Investigation of Details in the Transition to Synchronization in Complex Networks by Using Recurrence Analysis

Roberto C. Budzinski *, Bruno Rafael Reichert Boaretto, Thiago L. Prado and Sergio Roberto Lopes

Departamento de Física, Universidade Federal do Paraná, 81531-980 Curitiba, PR, Brazil; brunorafaelrboaretto@gmail.com (B.R.R.B.); thiagolprado@gmail.com (T.L.P.); sergio.roberto.lopes@gmail.com (S.R.L.)

* Correspondence: roberto.budzinski@gmail.com

Received: 18 March 2019; Accepted: 17 April 2019; Published: 20 April 2019

Abstract: The study of synchronization in complex networks is useful for understanding a variety of systems, including neural systems. However, the properties of the transition to synchronization are still not well known. In this work, we analyze the details of the transition to synchronization in complex networks composed of bursting oscillators under small-world and scale-free topologies using recurrence quantification analysis, specifically the determinism. We demonstrate the existence of non-stationarity states in the transition region. In the small-world network, the transition region denounces the existence of two-state intermittency.

Keywords: neural network; synchronization; nonlinear dynamics

1. Introduction

Many natural phenomena can be modeled and studied through a mathematical approach. Especially, complex networks are useful for analyzing problems involving physical, biological, chemical, engineering, and even social perspectives [1,2]. In this way, coupled oscillators are able to investigate a large class of dynamical systems in a theoretical, computational or even experimental field [3–6]. As an example, neural networks can be understood as coupled oscillators where it is possible to associate the bursting neuron to a phase oscillator [7].

In the scenario of complex networks, the neural system can be modeled on two scales. In the internal connection scheme, each network node can be understood as a neuron and their connections as the edges [8], which are able to simulate a single network, as used in many works [9–13]. On the other hand, considering the inter-networks connection scheme, it is possible to consider a neural system composed of different sub-areas, so each sub-network can be understood as a node and their connections as the edges, building a network of networks [14–18].

The connection architecture of complex networks is very important in the dynamical properties observed at a global level of behavior. Regarding neural networks, many topologies are considered, such as small-world, scale-free, and random ones [9,16,19] where a transition from unsynchronized to synchronized states is observed. The role of connection architecture is very important to the paths to synchronization, where different phenomena can be observed as a function of topology [20–22]. In real neural systems, the characteristics of these topology schemes are observed [23–25], which motivates the investigation of the influence of topology on the dynamical properties of the networks.

Regarding complex networks, it is known that this kind of system can show emergent behavior, where the global behavior observed is richer than the sum of the individual element behaviors. In this way, the existence of non-monotonic transitions to synchronization as a function of coupling strength in neural networks [26–30], where non-stationary states can be noticed, has been reported in the literature. In some cases, in the transition region, on–off intermittency in the two states has been observed, where the network displays the existence of two locally stable states but globally unstable ones [18,26], as defined in [31]. In [9], the dynamical properties regarding synchronization and transition characteristics are studied as a function of the connection architecture, with both small-world and random topologies being considered. In the present paper, we extend the analysis and consider the scale-free topology, since there are topological differences between the connection schema. In the small-world network, there is a local connection structure plus a non-local one. On the other hand, in the random network, there is homogeneity in the connectivity distribution, which consists of an assortative network and composes a very different scenario in comparison to the scale-free network, which generally consists of hubs of connections and forms a disassortative network [32,33]. In fact, these differences may influence the dynamical properties of systems regarding synchronization, as observed in [20–22,33].

In order to simulate the neural behavior, we consider the coupled map developed by Rulkov [34]. Using this model, it is possible to reproduce bursting behavior, which is characterized by a sequence of chaotic spikes followed by a period of resting [35]. This kind of neural activity is observed in real neural systems, as reported in [36–38]. The building of networks involves two different topologies: small-world, obtained through the Newman–Watts route [39], and scale-free (power law distribution of connectivity), obtained through the Barabasi–Albert approach [40], since these topologies characteristics are observed in real neural systems.

To perform the numerical analyses of dynamical properties regarding phase synchronization and (non-)stationarity of the transition region, the Kuramoto order parameter [3] and recurrence quantification analysis (RQA) [41,42] are used. In RQA, the determinism is used, which evaluates the ratio of recurrent points that belong to diagonal structures. To use the Kuramoto order parameter, data from each neuron that composes the network is necessary since a phase is associated with the bursting behavior of all neurons. The use of recurrence quantification only requires a time series that characterizes the dynamical systems. In the case of networks, the determinism is evaluated from the mean field time series. It is known that the mean field of a phase synchronized network has a "periodic" oscillation where the amplitude is bigger than in an unsynhcronized case, as observed in [11,26]. This approach makes an experimental validation possible since the recurrence quantification analysis is able to analyze experimental time series [43–46].

In this paper, we investigate a network composed of 1024 chaotic bursting neurons simulated through the Rulkov map in terms of phase synchronization and transition region characteristics as a function of coupling strength. Here, we focus on the small-world and scale-free topologies and their influence on the dynamical properties of the neural networks. We show that the transition to phase synchronization depicts non-stationary characteristics; however, the transition occurs for smaller values of coupling strength in the scale-free network. On the other hand, the existence of two-state on–off intermittency is observed for the small-world network.

The paper is divided as follows. In Section 2, the Rulkov map and the bursting behavior are described. In Section 3, the connection architectures are shown, and the small-world and scale-free building approaches are described. In Section 4, the synchronization and intermittency quantifiers are presented. In Section 5, the results and a discussion are presented, which support the conclusions in Section 6.

2. Rulkov Map

The Rulkov map is able to reproduce different dynamical behaviors, such as chaotic spikes and a set of chaotic bursts [47]. The bi-dimensional Rulkov map can be described as

$$x_{t+1,i} = \frac{\alpha}{1 + x_{t,i}^2} + y_{t,i} + I_{t,i}, \tag{1}$$

$$y_{t+1,i} = y_{t,i} - \beta x_{t,i} + \gamma, \tag{2}$$

where x_i and y_i are the fast and slow variables of the ith neuron. The set of parameters ($\alpha = 4.1$, $\beta = \gamma = 0.001$) is chosen in order to obtain the bursting behavior, following [34]. The coupling term I_i represents the influence of the other neurons in the ith neuron and is given by

$$I_{t,i} = \frac{\varepsilon}{\chi} \sum_{j=1}^{N} a_{i,j} x_{t,j}, \tag{3}$$

where ε is the coupling strength, χ is the normalization factor given by the average number of connections in the network [48], and $a_{i,j}$ is the adjacency matrix element where

$$a_{ij} = \begin{cases} 1, & \text{if } i \text{ and } j \text{ are connected,} \\ 0, & \text{if } i \text{ and } j \text{ are not connected.} \end{cases}$$

Here, we consider a small-world and a scale-free networks.

Figure 1 depicts an example of the dynamical behavior of the fast and slow variables of the Rulkov map. Panel (**a**) represents x where bursting behavior can be observed, which is characterized by a sequence of chaotic spikes followed by a resting time. Panel (**b**) depicts y as a function of t, where each time that a burst starts, the slow variable assumes a maximum. For the interval of coupling strength (ε) used, the bursting behavior is maintained.

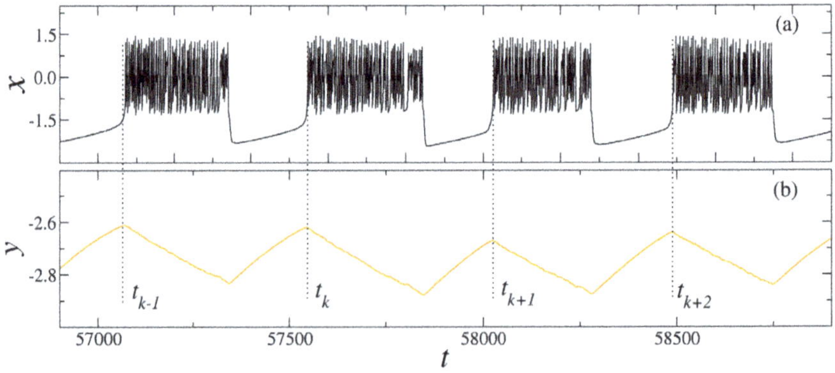

Figure 1. Dynamical behavior of an oscillator obtained through a Rulkov map. Here, panel (**a**) depicts the fast variable of the system (x), which can be understood as the membrane potential of the neuron. Panel (**b**) depicts the slow variable of the system (y) and is useful for evaluating the oscillator phase since y assumes a maximum at every burst start.

Since the fast variable x reproduces the bursting behavior, it is possible to understand x as the neuron membrane potential, and the mean field potential can be described as [9]

$$\overline{V}_t = \frac{1}{N} \sum_{i=1}^{N} x_{t,i}, \tag{4}$$

where $N = 1024$ is the number of oscillators (neurons) in the network.

3. Connection Architecture

The topology is very important to the dynamical properties of the networks. In fact, in neural networks, the connection scheme can be even more important than the structural variants of the neurons [49]. In this way, many results show that neural systems can display different topological properties, with small-world and scale-free characteristics being observed in real systems [50–53].

To build the scale-free network, we used the Barabasi–Albert approach [54], which can be described by two processes:

- the network is expanded by the addition of new nodes;
- the nodes added preferentially build connections to nodes that are already well-connected.

This approach is able to build a network where the distribution of the connectivity probability follows a power law: $P(k) \sim k^{-\nu}$ with $\nu > 1$.

Regarding real neural systems, experimental evidence indicates that some brain activities may depict scale-free properties where the average number of connections is ≈ 4 [52,53]. Moreover, it is known that the human functional network may present scale-free characteristics where there is a power-law connectivity distribution following an exponent $2.0 \leq \nu \leq 2.2$ [52].

The scale-free network used in this paper has $N = 1024$ nodes with $n = 4088$ connections and $\nu = 2.20$, which gives an average number of connections of $\chi_{\rm sf} = 4$, which is used as the normalization factor in Equation (3).

In order to obtain the small-world connection matrix, the Newman–Watts route is used [39], where non-local connections are added in a second neighborhood regular network with a probability of p. The network is composed of $N = 1024$ nodes, and the total number of connections is given by

$$n = \underbrace{4N}_{\text{local}} + \underbrace{N(N-5)p}_{\text{nonlocal}}, \tag{5}$$

where $p = 1$ leads to a globally connected network. Here, the number of local connection is 4096, and by using $p = 5.3 \times 10^{-4}$, 554 non-local connections are obtained, which leads to $n = 4650$ connections. For this network, the average number of connections is given by $\chi_{\rm sw} = 4.53$.

It is possible to evaluate the average path length (L) and the clustering coefficient (C) [50] of the small-world network. For the matrix used in this paper, $C_{\rm sw} = 0.3585 \sim 10^{-1}$ and $L_{\rm sw} = 6.012 \sim 1$ are obtained using the NetworkX library [55]. However, it is possible to evaluate these quantifiers for an equivalent (with a similar number of connections) random network using the expressions $L_{\rm random} \sim \ln(N)/\ln(n/N)$ and $C_{\rm random} \sim n/N^2$ [56].

Defining the merit variable $\Gamma = \lambda_C/\lambda_L$ where $\lambda_L = L_{\rm sw}/L_{\rm random}$ and $\lambda_C = C_{\rm sw}/C_{\rm random}$ [57,58] is possible to analyze the small-world existence condition. If $\Gamma > 1$, then the network considered has a small-world topology. For the network used here, $C_{\rm random} \sim 10^{-3}$ and $L_{\rm random} \sim 1$, which leads to $\Gamma \sim 10^2$, confirming that the networks used have a small-world topology.

Figure 2 depicts the graph representation of the networks used in this paper. Panel (**a**) shows the small-world network, and panel (**b**) represents the scale-free one. One difference between the networks consists of the degree of connectivity, since in the small-world network the most connected neuron has 9 connections, while in the scale-free case this number is bigger than 140. Besides this, in the scale-free case is observed the formation of hubs characterized by larger connected neurons, which leads to non-homogeneity in the connectivity degree, while in the small-world network a different scenario is noticed, and the degree of neurons connectivity is similar.

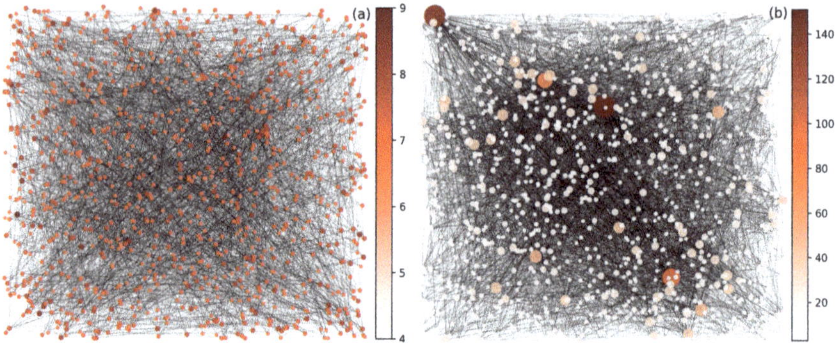

Figure 2. Representation of the networks used in this paper. Panel (**a**) depicts the small-world network with 1024 nodes and 4650 connections, and panel (**b**) the scale-free network with 1024 nodes and 4088 connections. The existence of hubs of connections is observed in the scale-free network, and the degree of connectivity decays in a power-law form, while in the small-world one the degree of connectivity is similar between nodes.

4. Quantifiers

The synchronization of chaotic systems has been extensively studied since the last century [50,59–61]. Here, we focus on the phase synchronization of networks composed of bursting neurons. There are different approaches to obtain the phase of chaotic oscillators; however, we have used the slow variable y to identify every time a burst starts. In this way, the phase is increased by 2π every maximum of y, and a continuous variation of the phase is obtained through [60,62]

$$\theta(t) = 2\pi \mathrm{k} + 2\pi \frac{t - t_{\mathrm{k}}}{t_{\mathrm{k}+1} - t_{\mathrm{k}}}, \qquad t_{\mathrm{k}} < t < t_{\mathrm{k}+1}, \tag{6}$$

where $t_{k,i}$ is the time when the kth burst of the ith neuron starts. It is important to emphasize that the bursting behavior is maintained for the entire interval of coupling strength used (ε), which means that phase can be obtained from the time series of the maximum of y without reconstruction of the dynamics in the higher dimensional phase space [60].

To evaluate phase synchronization based on phase θ, the Kuramoto order parameter is used [3]. Considering the phase of all neurons, it is possible to define

$$R(t) = \left| \frac{1}{N} \sum_{j=1}^{N} \mathrm{e}^{i\theta_j(t)} \right|. \tag{7}$$

If the system is (not) in a phase synchronized state, then $R \to 1$ ($R \to 0$).

To obtain the synchronization level as a function of the coupling strength, the time average of $R(t)$ can be computed for each value of ε. The mean value of the Kuramoto order parameter is given by

$$\langle R \rangle = \frac{1}{t_{\mathrm{f}} - t_0} \sum_{t=t_0}^{t_{\mathrm{f}}} R(t), \tag{8}$$

where t_{f} is the total simulation time, and t_0 is the transient time.

Recurrence quantification analyses are useful for analyzing the synchronization characteristics of dynamical systems [9,26,42,63,64]. The original work about recurrence analyses is based on a visual approach to identifying dynamical properties [65]. On the other hand, it is known that the mathematical structure of recurrence plots (RP) offers information about the dynamics of a system [41,42].

Here, we perform recurrence analyses based on the time series of the mean field of the network, described by Equation (4). In this case, the recurrence matrix is defined as

$$\mathbf{R}_{ab}(\delta) = \Theta(\delta - ||\mathbf{x}_a - \mathbf{x}_b||), \mathbf{x}_a \in \mathbb{R},\ a, b = 1, 2, \cdots, S, \tag{9}$$

where Θ is the Heaviside function, δ is the recurrence threshold, and S is the size of time series analyzed. Based on a time series of size S, of the mean field potential of the network described by Equation (4), it is possible to obtain the recurrence matrix through the use of Equation (9). The procedure consists in analyzing whether each point of the time series is close enough to another point (the difference between them must be smaller or equal to δ). After considering the entire time series, a matrix $(S \times S)$ is obtained where if a state $\mathbf{x}_a$ is (not) recurrent to another $\mathbf{x}_b$, $\mathbf{R}_{ab}$ assumes one (zero). The visual approach to analyzing the system consists of relating the recurrent (not recurrent) point to a black (white) dot [65]. In order to fix $\delta \in [0, 1]$, the time series of the mean field must be normalized.

In order to analyze the synchronization of the network, the diagonal lines in the recurrence matrix are very important since their distributions are related to the regularity of the trajectories. A diagonal line of length ℓ is understood as a segment of the trajectory rather close to another segment of the trajectory during ℓ time steps in a different time [41].

In this way, the better quantifier for the synchronization analysis is the determinism, which gives the ratio of recurrent points that belong to diagonal lines over all recurrent points in the recurrence matrix defined by Equation (9). This method is useful since the mean field of a phase synchronized network, depicted by Equation (4), has "periodic" oscillations, as observed in [26,30].The determinism is defined as

$$\Delta(\ell_{\min}, \delta, \overline{V}) = \frac{\sum_{\ell=\ell_{\min}}^{S} \ell P(\ell, \delta)}{\sum_{\ell=1}^{S} \ell P(\ell, \delta)}, \tag{10}$$

where $\ell_{\min}$ is the minimum length to consider a diagonal line. $P(\ell, \delta)$ is the probability distribution function (PDF) of the diagonal lines.

In a similar approach that uses the Kuramoto order parameter, it is possible to evaluate the mean value of the determinism using

$$\langle \Delta \rangle = \frac{1}{t_f - t_0} \sum_{t=t_0}^{t_f} \Delta(t). \tag{11}$$

5. Results and Discussions

The numerical results have been obtained from initial conditions that follow a random distribution in the intervals described by the fast and slow variables (see Figure 1), which avoid any initial trend. The transient time is given by $t_0 = 100{,}000$.

An important point to note is the dependence of the recurrence quantification analyses (RQA) on the recurrence parameters, an particularly, the recurrence threshold (δ) and minimum diagonal length ($\ell_{\min}$). If $\delta \rightarrow 1$, all points will be considered recurrent, and the determinism becomes saturated [41]. Similar behavior is obtained if $\ell_{\min} \rightarrow 1$, since all points will be considered as diagonal structures [41]. In order to use RQA to investigate dynamical properties regarding phase synchronization and/or non-stationarity, it is necessary to optimize the quantifiers through the choice of recurrence parameters [9,43]. The choice of recurrence threshold parameter follows [43],where the condition is described by $d[\Delta(\delta)]/d\delta$ = "a maximum" leading to $\delta = 0.11$, which results in the highest sensitivity of the quantifier determinism(Δ). The minimum diagonal length ($\ell_{\min} = 35$) is chosen in order avoid small diagonals, and the determinism (Δ) is better able to distinguish the phase synchronized states [9]. The determinism is evaluated using a moving window of 10,000 points.

The phase synchronization main scenario is depicted in Figure 3, where the synchronization characteristics are studied as a function of the coupling strength. Here, the total time of simulation is given by $t_f = 250{,}000$. Panels (**a**) and (**b**) depict the mean value of the Kuramoto order parameter ($\langle R \rangle$) and the mean value of the determinism ($\langle \Delta \rangle$) for the scale-free network. Panels (**c**) and (**d**) depict the same analysis for the small-world network. Here, 20 different seeds for the initialization of the system are considered, and the vertical magenta bars indicate the dispersion (standard deviation) over initial conditions. For both networks, a clear transition from unsynchronized to phase synchronized state is observed; however, in the scale-free network, the transition occurs for smaller values of coupling strength in comparison to the small-world network. The dispersion over initial conditions is bigger at the transition region for both cases; however, for the small-world network, the phenomenon is more visible. A similar scenario is observed in [9].

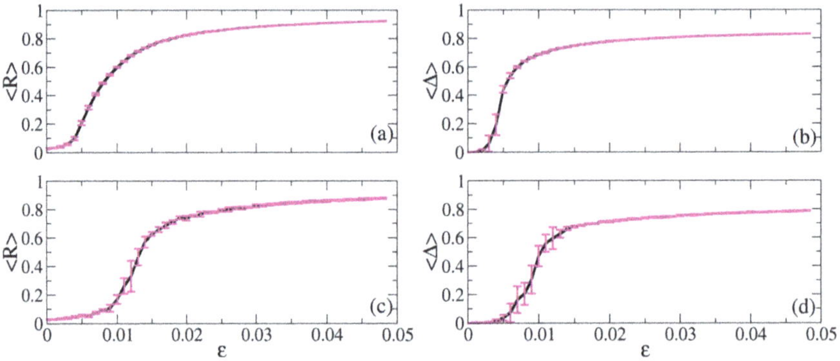

Figure 3. Panels (**a**) and (**c**) depict the mean value of the Kuramoto order parameter ($\langle R \rangle$) as a function of the coupling strength (ε) for the scale-free and small-world networks, respectively. Panels (**b**) and (**d**) depict the mean value of the determinism ($\langle \Delta \rangle$) for the same scale-free and small-world networks. A clear transition to phase synchronized states is observed in both networks.

In the scale-free network, the transition occurs where $0.0025 < \varepsilon < 0.012$, where the increase in the coupling strength makes the system reach the phase synchronized state. In the small-world network, the transition region is characterized by $0.005 < \varepsilon < 0.017$, and phase synchronization is observed for higher values of ε. This fact indicates that the scale-free network reaches the phase synchronized states for smaller values of ε in comparison to the small-world network. A similar scenario is observed in [66], where small-world networks reach synchronous behavior slower than other topologies. However, it is important to emphasize that the coupling term given by Equation (3) depends on the coupling strength (ε) and normalization factor χ. In this case, the normalization factor for each network is given by $\chi_{sf} = 4$ (scale-free) and $\chi_{sw} = 4.53$ (small-world), which leads to a relative factor of 1.13 between them. On the other hand, the difference in the coupling strength regime for the occurrence of the transition and phase synchronization is bigger than 1.13, so it is possible to conclude that the scale-free network reaches the phase synchronized state for a smaller coupling strength than the small-world one. Considering the small-world and random networks, it is possible to observe a higher level of synchronization in the random case [9,16]. Here, a similar scenario to the small-world and scale-free case is obtained; however, a new phenomenon regarding the critical value of ε is observed, as previously mentioned. In fact, the scale-free network presents a smaller value of ε to reach synchronization in comparison to the small-world and random cases, as studied in [9]. A similar scenario is observed in [20].

In order to investigate the details regarding synchronization characteristics, Figure 4 depicts the fast variable x for all neurons in the network as a function of t. Panel (**a**) represents the scale-free network with $\varepsilon = 0.002$, panel (**b**) $\varepsilon = 0.005$, and panel (**c**) $\varepsilon = 0.040$. Panel (**d**) represents the small-world network with $\varepsilon = 0.002$, panel (**e**) $\varepsilon = 0.011$, and panel (**f**) $\varepsilon = 0.040$. In the unsynchronized states (panels (a) and (d)), the neurons start their bursts without any coherence; however, in the transition region (panels (**b**) and (**e**)), the formation of horizontal structures is noticed. Finally, in panels (**c**) and (**f**), the phase synchronization where the horizontal structures denounce the spatial–temporal coherence of their bursts can be noticed.

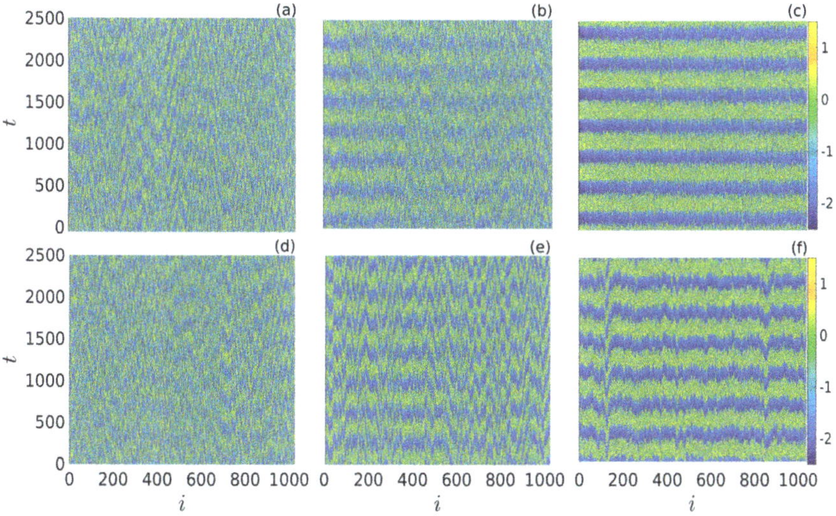

Figure 4. Spatial–temporal pattern of the membrane potential represented by the fast variable x for the scale-free network in panel (**a**) ($\varepsilon = 0.002$), panel (**b**) ($\varepsilon = 0.005$), and panel (**c**) ($\varepsilon = 0.040$) and for the small-world network in panel (**d**) ($\varepsilon = 0.002$), panel (**e**) ($\varepsilon = 0.011$), and panel (**f**) ($\varepsilon = 0.040$). In the phase synchronized states, horizontal lines, which are related to the bursting synchronization, are observed.

The mean field of the networks (Equation (4)) is depicted in Figure 5. It is important to note that the mean field consists of more easily experimentally accessible data than the individual signal. As observed in [12,30,67], the amplitude of the mean field of the network is related to the phase synchronization. Here, the scale-free (panels (**a**), (**b**), and (**c**)) and small-world networks (panels (**d**), (**e**), and (**f**)) are considered. In the unsynchronized cases (panels (**a**) and (**d**)) where $\varepsilon = 0.002$, the amplitude of the mean field $\overline{V}$ is vanishing. However, increases in the coupling strength (panel (**b**)—$\varepsilon = 0.005$ and panel (**e**)—$\varepsilon = 0.011$) lead the network to the transition region, where a small amplitude of $\overline{V}$ is observed. Importantly, the amplitude of the mean field varies as a function of t, as observed in [26], which indicates the possible existence of intermittency. Finally, for higher values of coupling $\varepsilon = 0.040$ (panels (**c**), and (**f**)), the mean field depicts an oscillatory behavior with a higher amplitude, and the frequencies are related to the bursting activity, which indicates phase synchronization behavior.

The "periodic" behavior of the mean field time series makes possible the use of RQA to evaluate the synchronization characteristics of the networks [9,26,67]. As depicted in Figure 3, the mean value of the determinism ($\langle\Delta\rangle$) is able to distinguish the level of phase synchronization. Figure 6 depicts the recurrence plot obtained from the recurrence matrix described by Equation (9), where the black dots are related to the recurrent points and the white dots to the non-recurrent ones. Here, the recurrence matrix is obtained from the mean field time series described by Equation (4) and depicted in Figure 5. Panel (**a**) of Figure 6 depicts the recurrence plot of an unsynchronized state, while panel (**b**) depicts the recurrence plot of a phase synchronized one. It is noticed that in the phase synchronized case, the diagonal structures are much clearer, which leads to a higher value of determinism (Δ) and makes its use possible for analyzing the synchronization characteristics of networks.

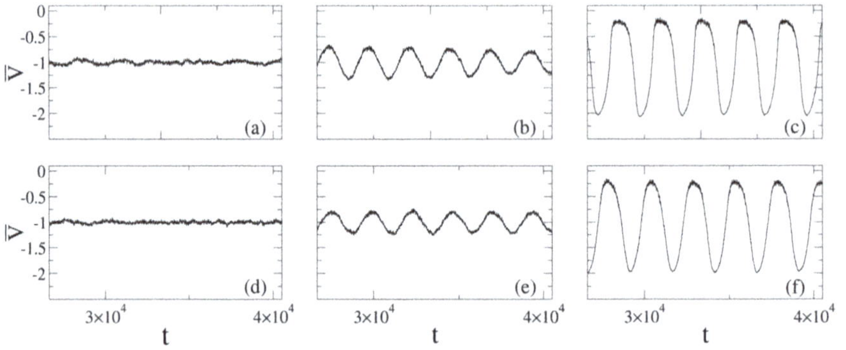

Figure 5. Mean field of the network described by Equation (4) for the scale-free network in panels (**a**), (**b**), and (**c**) and the small-world network in panels (**d**), (**e**), and (**f**). Here, the coupling values considered are the same as in Figure 4, where (**a**) and (**d**)—$\varepsilon = 0.002$, (**b**)—$\varepsilon = 0.005$, (**c**), and (**f**)—$\varepsilon = 0.040$ and (**e**)—$\varepsilon = 0.011$. The phase synchronized states (**c**) and (**f**) depict the higher amplitude of oscillations related to the bursting frequency.

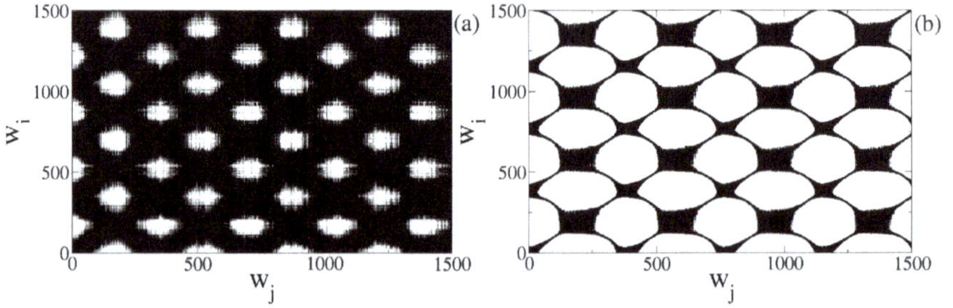

Figure 6. Recurrence plot obtained from the mean field of the network through Equation (9) for an unsynchronized case (panel (**a**)) and a synchronized one (panel (**b**)) where diagonal structures on the RP are observed, which leads to a higher value of determinism (Δ) in panel (**b**) than in panel (**a**).

The determinism is able to distinguish between more (less) synchronized states with just the use of the mean field time series [26]. In this way, it is possible to analyze the time series of the determinism in order to obtain information about synchronization characteristics as a function of t. Following [9], the temporal standard deviation of the determinism time series for different values of coupling strength (ε) offers information about the intermittency between states with different synchronization levels.

$$\sigma(A) = \sqrt{\frac{1}{T}\sum_{t=1}^{T}(A(t) - \langle A \rangle)^2}, \tag{12}$$

where A represents the Kuramoto order parameter (R) or the determinism (Δ), and T is the length of the time series considered. Here, a stationary time series of R or Δ leads to a vanishing value of σ. On the other hand, if the time series of R or Δ display an intermittent behavior, where the network assumes different synchronization level states as a function of t, then σ assumes higher values [18].

Figure 7 depicts the temporal standard deviation of the Kuramoto order parameter and the determinism time series as a function of the coupling strength (ε). Here, panels (**a**) and (**b**) depict $\sigma(R)$ and $\sigma(\Delta)$ for the scale-free network, while panels (**c**) and (**d**) depict $\sigma(R)$ and $\sigma(\Delta)$ for the small-world network, where σ is normalized by its maximum value. For all cases, it is observed that increases in the coupling strength lead to σ increases, until a maximum. After reaching the maximum, increases in ε lead to σ decreases until it reaches a vanishing value. Similar behavior is observed in [9] , where the maximum of σ is located at the transition region of the coupling value.Here, the observed behavior is the same since the transition region depicted in Figure 3 is related to the region where σ depicts higher values. This is indicative of intermittent behavior [18]. Again, for the scale-free network, the transition region, and consequently the intermittent region, occur at smaller values of coupling strength in comparison to the small-world network (see discussion about Figure 3). In particular, the maximum occurs at $\varepsilon \approx 0.0045$ in the scale-free network, and for the small-world network, the maximum is observed at $\varepsilon \approx 0.010$.

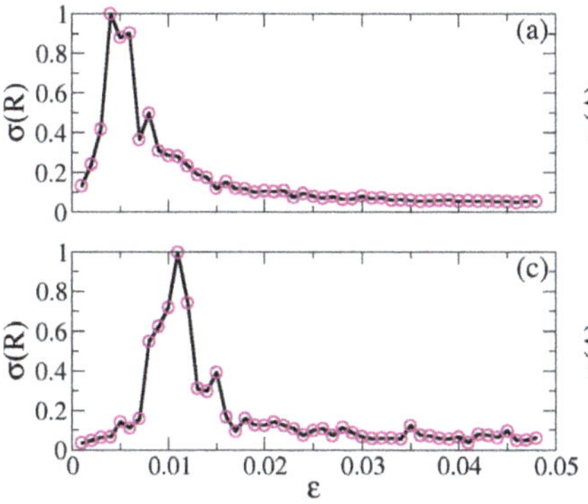

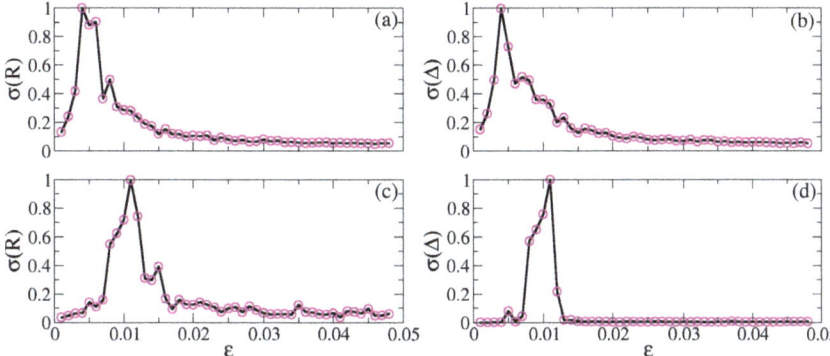

Figure 7. Normalized temporal standard deviation (σ), described by Equation (12), as a function of coupling strength for the scale-free network (panels (**a**) and (**b**)) and the small-world network (panels (**c**) and (**d**)). Here the time series of the Kuramoto order parameter (panels (**a**) and (**c**)) and the determinism time series (panels (**b**) and (**d**)) are considered. A higher value of σ indicates an intermittent behavior. The local maximum is observed in the transition region, indicating non-stationary transitions.

The determinism time series (Δ) is depicted in Figure 8 in order to obtain more details regarding the network dynamical properties in the different situations, as depicted in Figures 3–5. As previously mentioned, the recurrence matrix (Equation (9)) is evaluated from the mean field time series. In this way, the diagonal structures observed in Figure 6b are related to trajectories that are close enough (smaller or equal to δ) at different times. Through the optimization of δ following [43] , it is possible to use the determinism to quantify the synchronization level of the network as a function of time. In Figure 8, scale-free (panels (**a**), (**b**), (**c**), and (**d**)) and small-world networks (panels (**e**), (**f**), (**g**), and (**h**)) are considered, where panels (**a**) and (**e**) represent the unsynchronized states with $\varepsilon = 0.002$. It is observed that the determinism assumes small values, which correspond to a unsynchronized states [9]. Panels (**b**) and (**c**) represent the states at the transition region for the scale-free network ($\varepsilon = 0.0045$ and $\varepsilon = 0.0050$), where it is observed that the determinism shifts between different values. This fact indicates that the network assumes different synchronization states as a function of t, characterizing an intermittent behavior. Similarly, panels (**f**) and (**g**) represent the states at the transition region of the small-world network ($\varepsilon = 0.0105$ and $\varepsilon = 0.0110$). Again, an intermittent behavior is observed, where the determinism shifts between different values; however, in this case, two plateaus of synchronization are observed, indicating the possible existence of two-state on–off synchronization [18,26,68]. In [9], a similar system is considered under small-world and random topologies. Despite the random case not showing the intermittence between two states, it is possible to observe differences in comparison to the scale-free case, since in the scale-free case, the existence of the higher state is less pronounced. Finally, panels (**d**) and (**g**) represent phase synchronized states for the scale-free and small-world networks ($\varepsilon = 0.040$), respectively,where a stationary signal is observed for high values of Δ in both cases.

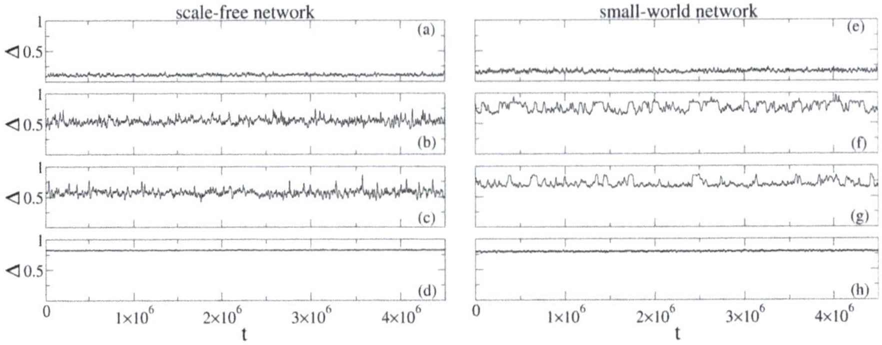

Figure 8. Examples of the determinism time series (Δ) evaluated from the mean field of the network for the scale-free network (panels (**a**), (**b**), (**c**), and (**d**)) and the small-world network (panels (**e**), (**f**), (**g**), and (**h**)). Here, panels (**a**) and (**c**) are representative of unsynchronized states ($\varepsilon = 0.002$), and panels (**b**), (**c**), (**f**), and (**g**) are representative of states at the transition region ($\varepsilon = 0.0045$, $\varepsilon = 0.0050$, $\varepsilon = 0.0105$ and $\varepsilon = 0.011$, respectively). At the transition region, it is observed that the system assumes different states of synchronization (different values of Δ).

Figure 9 depicts the probability distribution functions (PDFs) of the determinism time series with a length of 10^7 points, since long-term data is necessary to evaluate the temporal behavior of the networks [26]. The first row, panels (**a**), (**b**), (**c**), and (**d**) represent the scale-free network while the second row, panels (**e**), (**f**), (**g**), and (**h**) represent the small-world one. Panels (**a**) and (**e**) represent the unsynchronized state of scale-free and small-world networks, respectively, where $\varepsilon = 0.002$. In this case, a uni-modal distribution with a small dispersion over the mean value of Δ is observed, indicating the unsynchronized state [26].

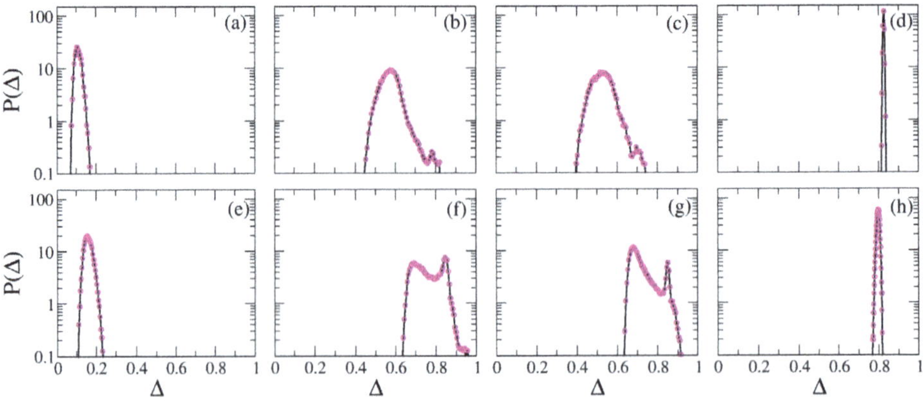

Figure 9. Probability distribution functions (PDFs) from the determinism time series (Δ) depicted in Figure 8. The panels follow the same scheme as in Figure 8, where the first row refers to the scale-free network, and the second row refers to the small-world network. In the transition region, the distribution is not symmetric and indicates a non-stationary behavior; however, only in the small-world case is the existence of two states observed.

If the coupling strength is increased, the networks assume the transition states, where the temporal standard deviations (σ) of the determinism time series are higher for both topologies considered, thus an intermittent behavior is expected. In this way, panels (**b**), (**c**), (**f**), and (**g**) represent the states at the transition region for the scale-free and small-world networks. Here, in panel (**b**) $\varepsilon = 0.0045$ and in panel (**c**) $\varepsilon = 0.0050$, a non-symmetric distribution of $\Delta(t)$ is observed, which indicates a non-stationary situation. In panels (**f**) and (**g**), the PDFs of $\Delta(t)$ at the transition state of the small-world network is depicted, where $\varepsilon = 0.0105$ and $\varepsilon = 0.0110$, respectively. Here, a non-stationary behavior is again observed since there is a non-symmetric distribution. However, for these cases, the existence of two states with similar occurrence probabilities is observed in the presence of two peaks in the PDFs. A similar scenario is observed in [9,18,26]. In fact, in [9] , the distribution of Δ with two peaks is observed in the small-world case. In the random case, a main peak and a shoulder in the distribution of Δ is observed. A comparison between the random and scale-free topologies makes it possible to notice that in both cases, there is no two-state intermittence; however, the random case presents a more unsymmetric PDF of Δ, which indicates a higher level of non-stationarity.

Finally, panels (**d**) and (**h**) of Figure 9 are representative of phase synchronized global stable states for scale-free and small-world networks where $\varepsilon = 0.040$. The PDFs are uni-modal with a very small dispersion over the mean value, which indicates that the state is stationary and globally stable [26]. It is important to emphasize that for the scale-free case (panel (**d**)), the mean value of the PDF of Δ is higher than in the small-world case (panel (**g**)), which suggest a higher level of phase synchronization.

6. Conclusions

The characteristics of synchronization in complex networks are very interesting and can be used for application in many nature phenomena [1,2,42,60]. Recently, studies have identified a non-stationary transition in bursting neurons networks [11,18,26] where, in some cases, the existence of two-state on–off intermittency is observed.

In order to analyze neural networks, we have used the Kuramoto order parameter [3], which uses the individual signal of all neurons in the network through the association of a phase based on the bursting activity. Moreover, we used recurrence quantification analysis [41,42,65], which is based on a time series that characterize the dynamical system. To analyze the mean field time series, we used the determinism, which is the ratio of recurrent points that belong to a diagonal structure.

Here, we have focused on a scale-free network composed of 1024 bursting neurons (oscillators) with 4088 connections and a small-world network composed of 1024 bursting neurons coupled trough 4096 local connections and 554 non-local ones. We noted a clear transition from unsynchronized to phase synchronized states where non-stationary behavior was observed in both cases. A similar scenario has been observed in [9,11,26,27,69]. Despite the higher number of connections in the small-world case, the transition region and the phase synchronized states were observed for smaller values of coupling strength in the scale–free case.

We have based our analyses in methodology proposed in [9], but here, we extended the study to a scale-free connection architecture since topology plays an important role in the dynamics of systems. It is important to emphasize that there are significant differences between small-world, random, and scale-free topologies. The small-world connection architecture is characterized by a regular local connections scheme with the addition of non-local ones. The random connection architecture is characterized by a more homogeneity in the connectivity, while the scale-free one is the opposite, since there are hubs of connections. The results have shown that the small-world case demonstrates the existence of two-state intermittence, as explored in [68]. Regarding the scale-free network, despite the non-stationary transition, this phenomenon is not observed. Similar results are obtained in [9] for random networks; however, it is also possible to observe differences between scale-free and random networks, as the random ones present a more non-stationary transition.

The study of complex networks has many application fields and is of great theoretical interest. Here, we have demonstrated that small-world and scale-free networks composed of bursting neurons can show non-stationary behavior at the transition to phase synchronization. These kinds of topologies find support in real neural systems, besides a large class of dynamical systems [1,23,50,52]. Thus, the present paper offers a theoretical/computational approach to understanding neural phenomena, since synchronization and intermittent behavior may be related to neural diseases [70,71] such as Parkinson's disease, autism, and Alzheimer's [72–75].

Author Contributions: Conceptualizon: R.C.B., B.R.R.B., T.L.P., S.R.L.; Software: R.C.B., B.R.R.B., T.L.P.; Numerical simulations: R.C.B.; Discussions: R.C.B., B.R.R.B., T.L.P., S.R.L.; Draft preparation: R.C.B.; Writing—review: R.C.B., B.R.R.B., T.L.P., S.R.L.; Supervision: S.R.L.

Acknowledgments: This study was financed in part by the Coordenação de Aperfeiçoamento de Pessoal de Nível Superior - Brasil (CAPES) - Finance Code 001. The authors acknowledge the support of Conselho Nacional de Desenvolvimento Científico e Tecnológico, CNPq - Brazil, grant number 302785/2017-5, Coordenação de Aperfeiçoamento de Pessoal de Nível Superior, CAPES, through project numbers 88881.119252/2016-01, and Financiadora de Estudos e Projetos (FINEP).

Conflicts of Interest: The authors declare no conflict of interest.

References

1. Strogatz, S.H. Exploring complex networks. *Nature* **2001**, *410*, 268–276. [CrossRef]
2. Bullmore, E.; Sporns, O. Complex brain networks: Graph theoretical analysis of structural and functional systems. *Nat. Rev. Neurosci.* **2009**, *10*, 186. [CrossRef] [PubMed]
3. Kuramoto, Y. *Chemical Oscillations, Waves, and Turbulence*; Springer Science & Business Media: Berlin/Heidelberg, Germany, 2012.
4. Rubido, N.; Cabeza, C.; Kahan, S.; Ávila, G.R.; Marti, A.C. Synchronization regions of two pulse-coupled electronic piecewise linear oscillators. *Eur. Phys. J. D* **2011**, *62*, 51–56. [CrossRef]
5. Ottino-Löffler, B.; Strogatz, S.H. Kuramoto model with uniformly spaced frequencies: Finite-N asymptotics of the locking threshold. *Phys. Rev. E* **2016**, *93*, 062220. [CrossRef]
6. Etémé, A.S.; Tabi, C.B.; Mohamadou, A. Firing and synchronization modes in neural network under magnetic stimulation. *Commun. Nonlinear Sci. Numer. Simul.* **2019**, *72*, 432–440. [CrossRef]
7. Ferrari, F.A.S.; Viana, R.L. Building phase synchronization equivalence between coupled bursting neurons and phase oscillators. *J. Phys. Commun.* **2018**, *2*, 025014. [CrossRef]
8. Hilgetag, C.C.; Kaiser, M. *Lectures in Supercomputational Neuroscience: Dynamics in Complex Brain Networks*; Springer: Berlin, Germany, 2008.
9. Budzinski, R.C.; Boaretto, B.R.R.; Rossi, K.L.; Prado, T.L.; Kurths, J.; Lopes, S.R. Nonstationary transition to phase synchronization of neural networks induced by the coupling architecture. *Phys. A Stat. Mech. Appl.* **2018**, *507*, 321–334. [CrossRef]
10. Batista, C.A.S.; Viana, R.L.; Ferrari, F.A.S.; Lopes, S.R.; Batista, A.M.; Coninck, J.C.P. Control of bursting synchronization in networks of Hodgkin-Huxley-type neurons with chemical synapses. *Phys. Rev. E* **2013**, *87*, 042713. [CrossRef] [PubMed]
11. Boaretto, B.R.R.; Budzinski, R.C.; Prado, T.L.; Kurths, J.; Lopes, S.R. Suppression of anomalous synchronization and nonstationary behavior of neural network under small-world topology. *Phys. A Stat. Mech. Appl.* **2018**, *497*, 126–138. [CrossRef]
12. Batista, C.A.S.; Lopes, S.R.; Viana, R.L.; Batista, A.M. Delayed feedback control of bursting synchronization in a scale-free neuronal network. *Neural Netw.* **2010**, *23*, 114–124. [CrossRef] [PubMed]
13. Yu, H.; Wang, J.; Liu, Q.; Deng, B.; Wei, X. Delayed feedback control of bursting synchronization in small-world neuronal networks. *Neurocomputing* **2013**, *99*, 178–187. [CrossRef]
14. Ferrari, F.A.S.; Viana, R.L.; Reis, A.S.; Iarosz, K.C.; Caldas, I.L.; Batista, A.M. A network of networks model to study phase synchronization using structural connection matrix of human brain. *Phys. A Stat. Mech. Appl.* **2018**, *496*, 162–170. [CrossRef]
15. Protachevicz, P.R.; Borges, R.R.; da Silva Reis, A.; Borges, F.S.; Iarosz, K.C.; Caldas, I.L.; Lameu, E.L.; Macau, E.E.N.; Viana, R.L.; Sokolov, I.M.; et al. Synchronous behaviour in network model based on human cortico-cortical connections. *Physiol. Meas.* **2018**, *39*, 7. [CrossRef]
16. Yamamoto, H.; Kubota, S.; Shimizu, F.A.; Hirano-Iwata, A.; Niwano, M. Effective subnetwork topology for synchronizing interconnected networks of coupled phase oscillators. *Front. Comput. Neurosci.* **2018**, *12*, 17. [CrossRef]
17. Mugnaine, M.; Reis, A.S.; Borges, F.S.; Borges, R.R.; Ferrari, F.A.S.; Iarosz, K.C.; Caldas, I.L.; Lameu, E.L.; Viana, R.L.; Szezech, J.D.; et al. Delayed feedback control of phase synchronisation in a neuronal network model. *Eur. Phys. J. Spec. Top.* **2018**, *227*, 1151–1160. [CrossRef]

18. Budzinski, R.C.; Boaretto, B.R.R.; Prado, T.L.; Lopes, S.R. Phase synchronization and intermittent behavior in healthy and Alzheimer-affected human-brain-based neural network. *Phys. Rev. E* **2019**, *99*, 022402. doi:10.1103/PhysRevE.99.022402. [CrossRef] [PubMed]
19. Lameu, E.L.; Batista, C.A.S.; Batista, A.M.; Iarosz, K.; Viana, R.L.; Lopes, S.R.; Kurths, J. Suppression of bursting synchronization in clustered scale-free (rich-club) neuronal networks. *Chaos Interdiscip. J. Nonlinear Sci.* **2012**, *22*, 043149. [CrossRef] [PubMed]
20. Gómez-Gardenes, J.; Moreno, Y.; Arenas, A. Paths to synchronization on complex networks. *Phys. Rev. Lett.* **2007**, *98*, 034101. [CrossRef]
21. Zhang, X.; Boccaletti, S.; Guan, S.; Liu, Z. Explosive synchronization in adaptive and multilayer networks. *Phys. Rev. Lett.* **2015**, *114*, 038701. [CrossRef] [PubMed]
22. Zhang, X.; Zou, Y.; Boccaletti, S.; Liu, Z. Explosive synchronization as a process of explosive percolation in dynamical phase space. *Sci. Rep.* **2014**, *4*, 5200. [CrossRef]
23. Varshney, L.R.; Chen, B.L.; Paniagua, E.; Hall, D.H.; Chklovskii, D.B. Structural Properties of the Caenorhabditis elegans Neuronal Network. *PLOS Comput. Biol.* **2011**, *7*, e1001066. [CrossRef]
24. He, Y.; Chen, Z.J.; Evans, A.C. Small-world anatomical networks in the human brain revealed by cortical thickness from MRI. *Cereb. Cortex* **2007**, *17*, 2407–2419. [CrossRef] [PubMed]
25. Stam, C.J.; Van Straaten, E.C.W. The organization of physiological brain networks. *Clin. Neurophysiol.* **2012**, *123*, 1067–1087. [CrossRef]
26. Budzinski, R.C.; Boaretto, B.R.R.; Prado, T.L.; Lopes, S.R. Detection of nonstationary transition to synchronized states of a neural network using recurrence analyses. *Phys. Rev. E* **2017**, *96*, 012320. [CrossRef]
27. Xu, K.; Maidana, J.P.; Castro, S.; Orio, P. Synchronization transition in neuronal networks composed of chaotic or non-chaotic oscillators. *Sci. Rep.* **2018**, *8*, 8370. [CrossRef]
28. Blasius, B.; Montbrió, E.; Kurths, J. Anomalous phase synchronization in populations of nonidentical oscillators. *Phys. Rev. E* **2003**, *67*, 035204. [CrossRef]
29. Prado, T.L.; Lopes, S.R.; Batista, C.A.S.; Kurths, J.; Viana, R.L. Synchronization of bursting Hodgkin–Huxley-type neurons in clustered networks. *Phys. Rev. E* **2014**, *90*, 032818. doi:10.1103/PhysRevE.90.032818. [CrossRef]
30. Boaretto, B.R.R.; Budzinski, R.C.; Prado, T.L.; Kurths, J.; Lopes, S.R. Neuron dynamics variability and anomalous phase synchronization of neural networks. *Chaos Interdiscip. J. Nonlinear Sci.* **2018**, *28*, 106304. [CrossRef]
31. Galuzio, P.P.; Lopes, S.R.; Viana, R.L. Two-state on-off intermittency caused by unstable dimension variability in periodically forced drift waves. *Phys. Rev. E* **2011**, *84*, 056211. [CrossRef]
32. Newman, M.E.J. The structure and function of complex networks. *SIAM Rev.* **2003**, *45*, 167–256. [CrossRef]
33. Liu, W.; Wu, Y.; Xiao, J.; Zhan, M. Effects of frequency-degree correlation on synchronization transition in scale-free networks. *Europhys. Lett.* **2013**, *101*, 38002. [CrossRef]
34. Rulkov, N.F. Regularization of synchronized chaotic bursts. *Phys. Rev. Lett.* **2001**, *86*, 183. [CrossRef] [PubMed]
35. Coombes, S.; Bressloff, P.C. *Bursting: The Genesis of Rhythm in the Nervous System*; World Scientific: Singapore, 2005.
36. Feudel, U.; Neiman, A.; Pei, X.; Wojtenek, W.; Braun, H.A.; Huber, M.; Moss, F. Homoclinic bifurcation in a Hodgkin–Huxley model of thermally sensitive neurons. *Chaos Interdiscip. J. Nonlinear Sci.* **2000**, *10*, 231–239. [CrossRef]
37. Braun, H.A.; Dewald, M.; Schäfer, K.; Voigt, K.; Pei, X.; Dolan, K.; Moss, F. Low-dimensional dynamics in sensory biology 2: Facial cold receptors of the rat. *J. Comput. Neurosci.* **1999**, *7*, 17–32. [CrossRef] [PubMed]
38. Gerstner, W.; Kistler, W.M. *Spiking Neuron Models: Single Neurons, Populations, Plasticity*; Cambridge University Press: Cambridge, UK, 2002.
39. Newman, M.E.J.; Watts, D.J. Scaling and percolation in the small-world network model. *Phys. Rev. E* **1999**, *60*, 7332–7342. doi:10.1103/PhysRevE.60.7332. [CrossRef]
40. Barabási, A.; Albert, R.; Jeong, H. Scale-free characteristics of random networks: the topology of the world-wide web. *Phys. A Stat. Mech. Appl.* **2000**, *281*, 69–77. [CrossRef]
41. Marwan, N.; Romano, M.C.; Thiel, M.; Kurths, J. Recurrence plots for the analysis of complex systems. *Phys. Rep.* **2007**, *438*, 237–329. [CrossRef]

42. Zou, Y.; Donner, R.V.; Marwan, N.; Donges, J.F.; Kurths, J. Complex network approaches to nonlinear time series analysis. *Phys. Rep.* **2019**, *787*, 1–97. [CrossRef]
43. Prado, T.d.L.; dos Santos Lima, G.Z.; Lobão-Soares, B.; do Nascimento, G.C.; Corso, G.; Fontenele-Araujo, J.; Kurths, J.; Lopes, S.R. Optimizing the detection of nonstationary signals by using recurrence analysis. *Chaos Interdiscip. J. Nonlinear Sci.* **2018**, *28*, 085703. [CrossRef]
44. Corso, G.; Prado, T.D.L.; Lima, G.Z.S.; Kurths, J.; Lopes, S.R. Quantifying entropy using recurrence matrix microstates. *Chaos Interdiscip. J. Nonlinear Sci.* **2018**, *28*, 083108. [CrossRef]
45. Lima, G.Z.S.; Lopes, S.R.; Prado, T.L.; Lobao-Soares, B.; do Nascimento, G.C.; Fontenele-Araujo, J.; Corso, G. Predictability of arousal in mouse slow wave sleep by accelerometer data. *PLoS ONE* **2017**, *12*, e0176761. [CrossRef]
46. Neves, F.M.; Viana, R.L.; Pie, M.R. Recurrence analysis of ant activity patterns. *PLoS ONE* **2017**, *12*, e0185968. [CrossRef] [PubMed]
47. Rulkov, N.F. Modeling of spiking-bursting neural behavior using two-dimensional map. *Phys. Rev. E* **2002**, *65*, 041922. [CrossRef] [PubMed]
48. Borges, R.R.; Borges, F.S.; Lameu, E.L.; Batista, A.M.; Iarosz, K.C.; Caldas, I.L.; Viana, R.L.; Sanjuán, M.A.F. Effects of the spike timing-dependent plasticity on the synchronisation in a random Hodgkin–Huxley neuronal network. *Commun. Nonlinear Sci. Numer. Simul.* **2016**, *34*, 12–22. [CrossRef]
49. Kandel, E.R.; Schwartz, J.H.; Jessell, T.M.; Siegelbaum, S.A.; Hudspeth, A.J. *Principles of Neural Science*; McGraw-Hill: New York, NY, USA, 2000.
50. Watts, D.J.; Strogatz, S.H. Collective dynamics of 'small-world' networks. *Nature* **1998**, *393*, 440–442. [CrossRef]
51. Lo, C.Y.; Wang, P.N.; Chou, K.H.; Wang, J.; He, Y.; Lin, C.P. Diffusion tensor tractography reveals abnormal topological organization in structural cortical networks in Alzheimer's disease. *J. Neurosci.* **2010**, *30*, 16876–16885. [CrossRef]
52. Chialvo, D.R. Critical brain networks. *Phys. A Stat. Mech. Appl.* **2004**, *340*, 756–765. [CrossRef]
53. Eguiluz, V.M.; Chialvo, D.R.; Cecchi, G.A.; Baliki, M.; Apkarian, A.V. Scale-free brain functional networks. *Phys. Rev. Lett.* **2005**, *94*, 018102. [CrossRef]
54. Barabási, A.L.; Albert, R. Emergence of scaling in random networks. *Science* **1999**, *286*, 509–512.
55. Hagberg, A.; Schult, D.; Swart, P. Exploring network structure, dynamics, and function using NetworkX. In Proceedings of the 7th Python in Science Conference, Pasadena, CA, USA, 19–24 August 2008; pp. 11–16.
56. Boccara, N. *Modeling Complex Systems*; Springer Science & Business Media: New York, NY, USA, 2010.
57. Albert, R.; Barabási, A. Statistical mechanics of complex networks. *Rev. Mod. Phys.* **2002**, *74*, 47–97. doi:10.1103/RevModPhys.74.47. [CrossRef]
58. Sporns, O.; Zwi, J.D. The small world of the cerebral cortex. *Neuroinformatics* **2004**, *2*, 145–162. [CrossRef]
59. Rosenblum, M.G.; Pikovsky, A.S.; Kurths, J. Phase synchronization of chaotic oscillators. *Phys. Rev. Lett.* **1996**, *76*, 1804. [CrossRef] [PubMed]
60. Boccaletti, S.; Kurths, J.; Osipov, G.; Valladares, D.; Zhou, C. The synchronization of chaotic systems. *Phys. Rep.* **2002**, *366*, 1–101. [CrossRef]
61. Bar-Yam, Y. *Dynamics of Complex Systems*; Addison-Wesley: Reading, MA, USA, 1997.
62. Ivanchenko, M.V.; Osipov, G.V.; Shalfeev, V.D.; Kurths, J. Phase Synchronization in Ensembles of Bursting Oscillators. *Phys. Rev. Lett.* **2004**, *93*, 134101. [CrossRef] [PubMed]
63. Lameu, E.L.; Yanchuk, S.; Macau, E.E.N.; Borges, F.S.; Iarosz, K.C.; Caldas, I.L.; Protachevicz, P.R.; Borges, R.R.; Viana, R.L.; Szezech, J.D., Jr.; et al. Recurrence quantification analysis for the identification of burst phase synchronisation. *Chaos Interdiscip. J. Nonlinear Sci.* **2018**, *28*, 085701. [CrossRef]
64. Santos, M.S.; Szezech, J.D., Jr.; Batista, A.M.; Caldas, I.L.; Viana, R.L.; Lopes, S.R. Recurrence quantification analysis of chimera states. *Phys. Lett. A* **2015**, *379*, 2188–2192. [CrossRef]
65. Eckmann, J.P.; Kamphorst, S.O.; Ruelle, D. Recurrence plots of dynamical systems. *Europhys. Lett.* **1987**, *4*, 973. [CrossRef]
66. Grabow, C.; Hill, S.M.; Grosskinsky, S.; Timme, M. Do small worlds synchronize fastest? *Europhys. Lett.* **2010**, *90*, 48002. [CrossRef]
67. Budzinski, R.C.; Boaretto, B.R.R.; Prado, T.L.; Lopes, S.R. Synchronization domains in two coupled neural networks. *Commun. Nonlinear Sci. Numer. Simul.* **2019**, *75*, 140–151. [CrossRef]

68. Galuzio, P.P.; Lopes, S.R.; Viana, R.L. Two-State On-Off Intermittency and the Onset of Turbulence in a Spatiotemporally Chaotic System. *Phys. Rev. Lett.* **2010**, *105*, 055001. [CrossRef]
69. Shanahan, M. Metastable chimera states in community-structured oscillator networks. *Chaos Interdiscip. J. Nonlinear Sci.* **2010**, *20*, 013108. [CrossRef] [PubMed]
70. Uhlhaas, P.J.; Singer, W. Neural synchrony in brain disorders: relevance for cognitive dysfunctions and pathophysiology. *Neuron* **2006**, *52*, 155–168. [CrossRef] [PubMed]
71. Lasemidis, L.D.; Sackellares, J.C. Chaos Theory and Eilepsy. *The Neuroscientist* **1996**, *2*, 118–126. [CrossRef]
72. Galvan, A.; Wichmann, T. Pathophysiology of parkinsonism. *Clin. Neurophysiol.* **2008**, *119*, 1459–1474. [CrossRef]
73. Dinstein, I.; Pierce, K.; Eyler, L.; Solso, S.; Malach, R.; Behrmann, M.; Courchesne, E. Disrupted neural synchronization in toddlers with autism. *Neuron* **2011**, *70*, 1218–1225. [CrossRef] [PubMed]
74. Aron, L.; Yankner, B.A. Neural synchronization in Alzheimer's disease. *Nature* **2016**, *540*, 207–208. [CrossRef] [PubMed]
75. Iaccarino, H.F.; Singer, A.C.; Martorell, A.J.; Rudenko, A.; Gao, F.; Gillingham, T.Z.; Mathys, H.; Seo, J.; Kritskiy, O.; Abdurrob, F.; et al. Gamma frequency entrainment attenuates amyloid load and modifies microglia. *Nature* **2016**, *540*, 230–235. [CrossRef] [PubMed]

Mathematical and Computational Applications

MDPI

Article

Suppression of Phase Synchronization in Scale-Free Neural Networks Using External Pulsed Current Protocols

Bruno Rafael Reichert Boaretto *, Roberto C. Budzinski , Thiago L. Prado and Sergio Roberto Lopes

Departamento de Física, Universidade Federal do Paraná, Curitiba, PR 81531-980, Brazil; roberto.budzinski@gmail.com (R.C.B.); thiagolprado@gmail.com (T.L.P.); sergio.roberto.lopes@gmail.com (S.R.L.)

* Correspondence: brunorafaelrboaretto@gmail.com

Received: 22 March 2019; Accepted: 23 April 2019; Published: 24 April 2019

Abstract: The synchronization of neurons is fundamental for the functioning of the brain since its lack or excess may be related to neurological disorders, such as autism, Parkinson's and neuropathies such as epilepsy. In this way, the study of synchronization, as well as its suppression in coupled neurons systems, consists of an important multidisciplinary research field where there are still questions to be answered. Here, through mathematical modeling and numerical approach, we simulated a neural network composed of 5000 bursting neurons in a scale-free connection scheme where non-trivial synchronization phenomenon is observed. We proposed two different protocols to the suppression of phase synchronization, which is related to deep brain stimulation and delayed feedback control. Through an optimization process, it is possible to suppression the abnormal synchronization in the neural network.

Keywords: neural network; synchronization; suppression of synchronization

1. Introduction

For years, the study of synchronization has been important since this phenomenon has been observed in many different biological systems, e.g., firefly communities, pacemaker cells of the heart, and crickets that chirp in unison [1–5]. In addition, the complexity seen in the brain is directly related to the distinct activation patterns of the neurons, which can be understood as a synchronization phenomenon. Particularly, the synchronization of neurons is important since anomalous synchronization can disrupt the brain functioning, generating disorders, such as Parkinson's disease (PD) and autism [6–10].

A possible neurosurgical treatment for PD is called deep brain stimulation (DBS), which consists of the insertion of an electric probe that emits electromagnetic signals in a target brain area [9,11,12]. A more recently developed technique is the noninvasive DBS, which consists of temporally interfering electric fields [12] generated outside of the cranium. Despite its long history of use, it is still unclear how DBS works [9,10]. Some studies indicate that high-frequency DBS replaces pathological low-frequency network oscillations in the rat model of Parkinson's disease with a regularized pattern of neuronal firing [13], and there is evidence that the DBS releases the activity patterns of groups of cells in the subthalamic nucleus that present abnormal synchronization due to PD, which destroys neurons in basal ganglia [14]. Depending on the frequency of the signal, it allows suppressing the symptoms of Parkinson's disease [13,15].

A healthy human brain consists of $\sim 10^{11}$ interconnected neurons through $\sim 10^{15}$ synapses [16,17]. In the theoretical point of view, a possible way to study coupled neurons is given by the computational simulation of complex networks, where each site of the network corresponds to a neuron and its connections are represented by the edges of the network [18]. In this scenario, distinct topologies or connection architectures have been successfully used to simulate the interconnections of the human brain, such as small-world, scale-free and random topologies [18–23].

In this study, we simulated a neural network composed of $N = 5000$ neurons in a scale-free topology, where each neuron was modeled by a Hodgkin–Huxley-type model [24–26]. This model is characterized by the insertion of two temperature sensitive parameters, and two additional slow ionic currents to the original ideas of Hodgkin and Huxley [27], which can be understood as the contribution of calcium ion channels [28].

It is observed in the literature that a neural network under a small-world topology can display abnormal phase synchronization for weak coupling regime since the phase synchronization regime in this region is characterized by a non-monotonic evolution of synchronization levels as a function of the coupling between neurons [29–31]. In fact, this kind of behavior has also been observed in non-identical coupled Rössler oscillators [32]. Recently, the mechanism behind the abnormal synchronization in a neural network composed of bursting neurons is explored and the relationship between the individual neuron behavior and the network synchronization helps to understand the phenomenon [33,34]. In [33], it is shown that the occurring of abnormal synchronization is related to the periodic inter-burst interval of the uncoupled neuron. Besides that, in [34], it is observed that the abnormal synchronization occurs due to an interplay between the periodic individual behavior and the influence of coupling strength, which is strong enough to induce the network to phase synchronization without destroying the influence of individual periodic behavior.

Here, we extend the study of abnormal synchronization to the scale-free connection architecture, since the topology plays an important role in the dynamics of systems [35–37]. Scale-free topology is different from the small-world one since the scale-free scheme presents a high degree of heterogeneity where neurons with a high connectivity degree are connected with neurons with low connectivity degree [37,38]. Thus, we study the existence of abnormal synchronization in a scale-free neural network and its suppression by the application of a disturbance in the network neurons. This perturbation is characterized by the application of a pulsed external current, which can be described as a theoretical interpretation of the DBS treatment.

We considered another suppression strategy that consists of the reapplication of a fraction of the signal generated by the neurons, which is called delayed-feedback-control (DFC). It was first applied experimentally in vitro in a spontaneously bursting neural network [39,40] and is frequently used in neural stimulation treatments [41,42].

To quantify synchronization of the network, we used the order parameter proposed by Kuramoto [43], which is able to capture information about phase synchronization of the system using data of each neuron. In this sense, we show that the suppression methods proposed are able to suppress the anomalous synchronization without affecting the regular synchronized states, which occur for large values of the coupling.

The paper is organized as follows. In Section 2, we introduce the details of the connection scheme and the used neuronal model. In Section 3, we introduce the quantification of PS by using of the Kuramoto order parameter. In Section 4, details of the perturbation methods imposed into the system, as well as the results obtained with each perturbation are discussed. Our conclusions are in the last section.

2. Neural Model and Connection Scheme

We studied the dynamical behavior of a neural network composed of 5000 neurons connected in a scale-free topology. In this case, the number of connections per node presents a statistical power-law dependence $P(n) \sim n^{-\kappa}$ [44,45]. The values of the scaling exponent are within $2 \leq \kappa \leq 2.2$ and the average connection $\langle n \rangle \approx 4$ [21,41,46], where $P(n)dn$ is the probability to find a node with a degree in the interval from n to $n + dn$.

Scale-free networks can be obtained by the Barabasi–Albert procedure through a sequence of steps starting from an initial lattice with a small number N_0 of nodes randomly connected [44,45]. At each step, a new node is inserted in the network, which is randomly connected to $n \geq 2$ nodes. The process is repeated until the network reaches the desired number of nodes. In this work, we used $N = 5000$ nodes. To generate a scale-free network, we used the Python library *NetworkX* [47], which have us $\kappa \approx 2.2$.

To simulate the individual neuron dynamics, we used a Hodgkin–Huxley-type model [25,26], where the adaptation takes into account the addition of two slow ionic fluxes. Mathematically, the neuronal model used in this work describes the temporal dynamics of the neuron membrane potential as a function of the ionic fluxes. This adaptation also includes temperature sensitive parameters. The temporal evolution of the membrane potential V_i is described by

$$C\frac{dV_i}{dt} = -J_{i,\mathrm{Na}} - J_{i,\mathrm{K}} - J_{i,\mathrm{sd}} - J_{i,\mathrm{sa}} - J_{i,\mathrm{L}} + J_{i,\mathrm{coup}}, \tag{1}$$

where C is the specific membrane capacitance of neurons measured in μF/cm^2; V_i is measured in mV; $J_{i,\mathrm{Na}}$, $J_{i,\mathrm{K}}$, and $J_{i,\mathrm{L}}$ are the sodium, potassium and non-gated channels fluxes of the original Hodgkin–Huxley model [27], respectively, which are measured in μA/cm^2; and $J_{i,\mathrm{sd}}$ and $J_{i,\mathrm{sa}}$ are the two slow ionic fluxes added by Braun et al. to this model and the are associate to calcium flux [28].

The electrical fluxes related to the ion and leak channels are given by conductance-based expressions [25]

$$J_{i,\mathrm{Na}} = \rho \bar{g}_{\mathrm{Na}} \alpha_{i,\mathrm{Na}}(V_i - E_{\mathrm{Na}}), \tag{2}$$
$$J_{i,\mathrm{K}} = \rho \bar{g}_{\mathrm{K}} \alpha_{i,\mathrm{K}}(V_i - E_{\mathrm{K}}), \tag{3}$$
$$J_{i,\mathrm{sd}} = \rho \bar{g}_{\mathrm{sd}} \alpha_{i,\mathrm{sd}}(V_i - E_{\mathrm{sd}}), \tag{4}$$
$$J_{i,\mathrm{sa}} = \rho \bar{g}_{\mathrm{sa}} \alpha_{i,\mathrm{sa}}(V_i - E_{\mathrm{sa}}), \tag{5}$$
$$J_{i,\mathrm{L}} = \bar{g}_{\mathrm{L}}(V_i - E_{\mathrm{L}}), \tag{6}$$

where $\bar{g}_{\mathrm{Na}}$, $\bar{g}_{\mathrm{K}}$, $\bar{g}_{\mathrm{sd}}$, $\bar{g}_{\mathrm{sa}}$, and $\bar{g}_{\mathrm{L}}$ are the maximum (specific) conductances measured in mS/cm^2, and E_{Na}, E_{K}, E_{sd}, E_{sa}, and E_{L} denote the reversal Nernst potentials for each ionic current measured in mV. The term ρ refers to a temperature dependence of the model and it is described by

$$\rho = \rho_0^{(T-T_0)/\tau_0}, \tag{7}$$

where T, T_0 and τ_0 and ρ_0 are constants of the model.

The temporal evolution of the activation functions $\alpha_{i,\mathrm{Na}}$, $\alpha_{i,\mathrm{K}}$,$\alpha_{i,\mathrm{sd}}$, and $\alpha_{i,\mathrm{sa}}$ are described by

$$\frac{d\alpha_{i,\mathrm{Na}}}{dt} = \frac{\phi}{\tau_{\mathrm{Na}}}(\alpha_{i,\mathrm{Na},\infty} - \alpha_{i,\mathrm{Na}}), \tag{8}$$
$$\frac{d\alpha_{i,\mathrm{K}}}{dt} = \frac{\phi}{\tau_{\mathrm{K}}}(\alpha_{i,\mathrm{K},\infty} - \alpha_{i,\mathrm{K}}), \tag{9}$$
$$\frac{d\alpha_{i,\mathrm{sd}}}{dt} = \frac{\phi}{\tau_{\mathrm{sd}}}(\alpha_{i,\mathrm{sd},\infty} - \alpha_{i,\mathrm{sd}}), \tag{10}$$
$$\frac{d\alpha_{i,\mathrm{sa}}}{dt} = \frac{\phi}{\tau_{\mathrm{sa}}}(-\eta J_{i,\mathrm{sd}} - \gamma \alpha_{\mathrm{i,sa}}), \tag{11}$$

where τ_{Na}, τ_{K}, τ_{sd}, and τ_{sa} are constants [25]. The parameter η works to increase calcium ion concentration following $J_{i,\mathrm{sa}}$, while γ accounts for active the elimination of intracellular calcium [28]. ϕ is another temperature dependence of the model, given by

$$\phi = \phi_0^{(T-T_0)/\tau_0}. \tag{12}$$

The functions $\alpha_{i,Na,\infty}$, $\alpha_{i,K,\infty}$, and $\alpha_{i,sd,\infty}$ are described by

$$\alpha_{i,\mathrm{Na},\infty} = \frac{1}{1 + \exp[-s_{\mathrm{Na}}(V_i - V_{0\mathrm{Na}})]}, \tag{13}$$

$$\alpha_{i,\mathrm{K},\infty} = \frac{1}{1 + \exp[-s_{\mathrm{K}}(V_i - V_{0\mathrm{K}})]}, \tag{14}$$

$$\alpha_{i,\mathrm{sd},\infty} = \frac{1}{1 + \exp[-s_{\mathrm{sd}}(V_i - V_{0\mathrm{sd}})]}, \tag{15}$$

where s_{Na}, s_{K}, s_{sd}, $V_{0\mathrm{Na}}$, $V_{0\mathrm{K}}$, and $V_{0\mathrm{sd}}$ are constants whose values are given in Table 1 following Braun et al. [25].

The coupling term $J_{i,\mathrm{coup}}$, in Equation (1), is an excitatory chemical synapse, since the synapse does not occur directly. In this way, the ith postsynaptic neuron receives signals from presynaptic ones [48]

$$J_{i,\mathrm{coup}} = \frac{\varepsilon}{\langle n \rangle} \sum_{j=1}^{N} e_{i,j} r_j (V_{\mathrm{syn}} - V_i), \tag{16}$$

where ε is the coupling parameter that controls the coupling intensity. $\langle n \rangle$ is the normalization factor, defined as the average of connections number, which is $\langle n \rangle \approx 4$. V_{syn} is the synaptic reversal potential, set here as 20 mV, which assures that the contribution coming from the coupling is positive for all instant of time, characterizing an excitatory synapse. $e_{i,j}$ represents the elements of the adjacency matrix, which is a scale-free type. In this case, if the ith and jth neurons are connected, $e_{i,j} = 1$; otherwise, $e_{i,j} = 0$.

Added to the kinetic variable of the model, r_i refers to the fraction of bound receptors available to receive a connection. We used the equation of r_i proposed by Destexhe et al. [48],

$$\frac{dr_i}{dt} = \left(\frac{1}{\tau_r} - \frac{1}{\tau_d}\right) \frac{1 - r_i}{1 + \exp[-s_0(V_i - V_0)]} - \frac{r_i}{\tau_d}, \tag{17}$$

where s_0 is a unitary constant measured in (1/mV), $V_0 = -20$ mV, and $\tau_r = 0.5$ ms and $\tau_d = 8$ ms are constants associated to the rises and decays of the synaptic transition, respectively.

To integrate the set of coupled equations composing the model, we used Adams' predictor-corrector method [49] with an absolute tolerance less than 10^{-8}. Figure 1a depicts the typical membrane potential for a neuron, using the fixed set of parameter values shown in Table 1. As observed, the neuron depicts bursting dynamics characterized by a sequence of spikes followed by a resting time [50]. We refer to this dynamics as bursting regime and, throughout the coupling interval used here, the suppression process ensured that this regime is not lost, which makes possible the phase association and synchronization evaluation for all interval of coupling and suppression strength.

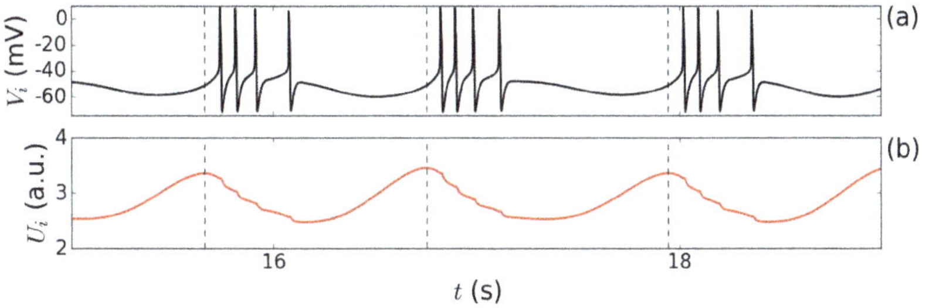

Figure 1. (**a**) Evolution of the dynamic behavior of the membrane potential V_i for the Hodgkin–Huxley-type model using the constants defined in Table 1. (**b**) The recovery variable $U_i \equiv 1/\alpha_{i,\mathrm{sa}}$ computed for each neuron, where the maximum of U_i corresponds to the beginning of each burst.

Table 1. Parameter values of the neuronal model according to Braun et al. [25].

Membrane Capacitance	$C = 1.0\ \mu\text{F}/\text{cm}^2$			
Characteristic Times (ms)	$\tau_{\text{Na}} = 0.05$	$\tau_{\text{K}} = 2.0$	$\tau_{\text{sd}} = 10$	$\tau_{\text{sa}} = 20$
Maximum Conductances (mS/cm^2)	$\bar{g}_{\text{Na}} = 1.5$ $\bar{g}_{\text{L}} = 0.1$	$\bar{g}_{\text{K}} = 2.0$	$\bar{g}_{\text{sd}} = 0.25$	$\bar{g}_{\text{sa}} = 0.4$
Reversal Potentials (mV)	$E_{\text{Na}} = 50$ $E_{\text{L}} = -60$	$E_{\text{K}} = -90$ $V_{0\text{Na}} = -25$	$E_{\text{sd}} = 50$ $V_{0\text{K}} = -25$	$E_{\text{sa}} = -90$ $V_{0\text{sd}} = -40$
Other Parameters	$\rho_0 = 1.3$ $s_{\text{Na}} = 0.25\ (1/\text{mV})$ $s_{\text{sd}} = 0.09\ (1/\text{mV})$	$\phi_0 = 3.0$ $\eta = 0.012\ \mu\text{A}$ $T = 13\ ^\circ\text{C}$	$T_0 = 25\ ^\circ\text{C}$ $\gamma = 0.17$	$\tau_0 = 10\ ^\circ\text{C}$ $s_{\text{K}} = 0.25\ (1/\text{mV})$

3. Phase Synchronization Quantifier

To quantify phase synchronization in the bursting regime, we associated a geometric phase to the sequence of bursts for each neuron. Figure 1b shows the auxiliary variable $U_i \equiv 1/\alpha_{i,\text{sa}}$ computed using Equation (11), where each maximum of U_i corresponds to the beginning of a burst of the ith neuron [51]. If $t_{k,i}$ is the beginning time of the kth burst of the ith neuron, the duration of the burst would be $t_{k+1,i} - t_{k,i}$, with $k = 0, 1, 2 \ldots$ and $i = 1, 2, \ldots, N$, consequently the phase would vary from $2\pi k$ to $2\pi(k+1)$, and it is defined for specific time t as [52]

$$\theta_i(t) = 2\pi k_i + 2\pi \frac{t - t_{k,i}}{t_{k+1,i} - t_{k,i}}, \qquad t_{k,i} \leq t < t_{k+1,i}. \tag{18}$$

Considering the geometric phase variable θ_i as defined in Equation (18), to quantify PS, we used the modulus of the Kuramoto order parameter $R(t)$ [53]

$$R(t) = \left| \frac{1}{N} \sum_{i=1}^{N} e^{i\theta_i(t)} \right|, \tag{19}$$

where $R(t)$ gives us a number between 0 (completely unsynchronized) and 1 (completely phase synchronized).

The order parameter oscillates in time for not fully synchronized neurons [43] and its temporal mean value is defined as

$$\langle R \rangle = \frac{1}{M} \sum_{j=1}^{M} R(t'_j), \tag{20}$$

being $t'_1 = t_i, t'_2 = t_i + h, \cdots, t'_M = t_f$, where $h = 0.01$, and t_i and t_f are initial and final times of the computation of $R(t)$.

To show the synchronization behavior of the network, Figure 2 depicts $\langle R \rangle$ as a function of the coupling parameter ε of a neural network given by Equations (1)–(17). To avoid any trend in the result, we used random initial conditions in the following intervals: $V_i \in [-65.0; 0.0]$; $\alpha_{\text{Na},i}, \alpha_{\text{K},i}, \alpha_{\text{sd},i}, \alpha_{\text{sa},i}$, $r_i \in [0.1; 1.0]$. Observe that, for $\varepsilon > \varepsilon^* = 0.007$ (mS/cm^2), the network followed a route of globally stable phase synchronized state, since $\langle R \rangle$ approached 1 as the coupling strength was increased. For the interval of coupling strength $0.002 < \varepsilon < 0.007$, the network exhibited a local maximum of phase synchronization ($\varepsilon \approx 0.004$). This behavior was also observed in small-world network [30,31,51], which characterized a non-monotonic evolution of the synchronization level as a function of the coupling that could be understood as an abnormal synchronization since PS occurred for a coupling $\varepsilon < \varepsilon^*$. In this way, it is known that several brain disorders, such as Parkinson's disease and autism, are related to abnormal neuronal synchronization [6–9], thus it is expected that the application of synchronization suppression methods may be useful to vanish the anomalous synchronization, as observed in Figure 2.

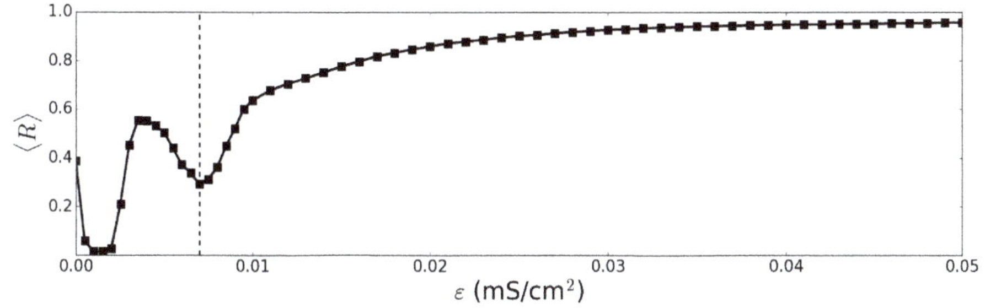

Figure 2. The mean order parameter $\langle R \rangle$ as a function of the coupling parameter ε for a scale-free neural network with 5000 identical neurons with randomly distributed initial conditions. The dashed vertical line represents the critical coupling ε^*.

4. Results and Discussions

Considering a scale-free neural network composed of $N = 5000$ identical neurons, we used the mean value of the Kuramoto order parameter $\langle R \rangle$ to evaluate the PS for different suppression synchronization protocols. Here, we used a transient time given by $t_i = 150$ s and the total simulation time was set to $t_f = 250$ s.

Motivated by experimental results [12,15], we made a perturbation in the network by applying an external pulsed current $\lambda(t)$ in Equation (1) of the neuronal model. Mathematically, the suppression method can be described as

$$\lambda(t) = \frac{\lambda_0}{2} + \sum_{m=1,3,\cdots}^{\infty} \frac{2\lambda_0}{m\pi} \sin\left(\frac{2m\pi t}{\tau}\right), \tag{21}$$

where λ_0 is the amplitude of each pulse measured in μA/cm^2, and τ is the period for which the current is successively turned on and off and measured in seconds. Figure 3 shows the evolution of $\lambda(t)/\lambda_0$ as a function of t/τ.

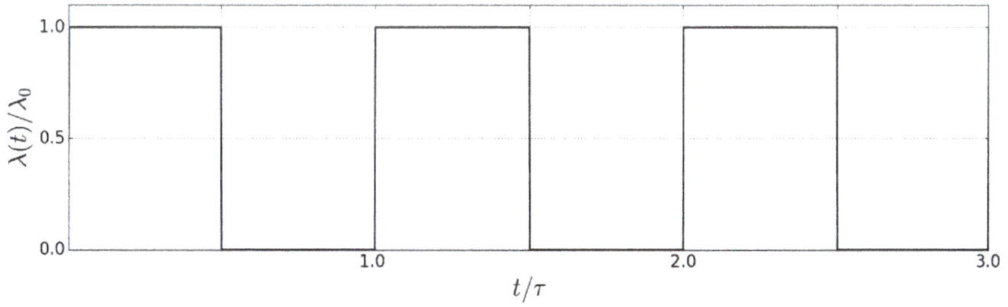

Figure 3. The on-off pulse evolution of $\lambda(t)/\lambda_0$ given by Equation (21).

Figure 4 depicts $\langle R \rangle$ as a function of the coupling ε for different values of amplitude λ_0. For amplitudes lower than $\lambda_0 < 0.05$ (μA/cm^2), the network still presented PS for small values of ε. However, for $\lambda_0 \geq 0.05$, the anomalous PS was suppressed without altering the globally stable state of synchronization for coupling value higher than the critical value ε^* (which remains constant in $\varepsilon^* \approx 0.007$ for $\lambda_0 < 0.2$). The frequency $\nu = 1/\tau$ was fixed at 140 Hz (which means $\tau \approx 0.0071$ s), since experimental results show that only a high frequency currents ($\nu > 100$ Hz [9,10,54]) could restore normal neural behavior in Parkinson's Disease (PD) [13].

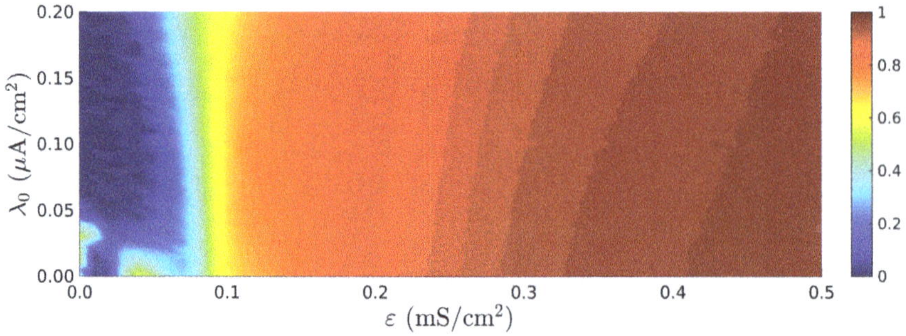

Figure 4. $\langle R \rangle$ as functions of ε and the amplitude of the external pulsed current λ_0, for a high frequency $\nu = 140$ Hz. For amplitudes $\lambda_0 \geq 0.05$, the method successfully suppressed the anomalous synchronization occurring for coupling strength $\varepsilon \lesssim 0.007$..

The next step consisted of the study of the heterogeneity of the scale-free network because one of the characteristics of the scale-free topology is the existence of *hubs*, which are characterized by neurons with high connectivity in the network [41,44,45]. Intuitively, it is believed that the perturbation presents greater influence when applied in the hubs since they have high connectivity in the network. We made a change in the applied current to apply the current in select groups of neurons

$$\lambda(t, \lambda_0, \tau) \rightarrow \lambda_i(t, \lambda_{0,i}, \tau)$$

where $\lambda_{0,i} = \lambda_0$ if $i \subset \mathcal{G}$ or $\lambda_{0,i} = 0$, otherwise $\mathcal{G}$ is a subset of neurons in the network. Here, it was studied how the PS varied for three different subsets $\mathcal{G}$. Firstly, we applied the current in the neurons with higher connectivity degree of the network, in this case, the order of $\mathcal{G}$ was $|\mathcal{G}| = N_{\text{hubs}}$. In the second case, the pulsed current was applied in random neurons of the network, and then $|\mathcal{G}| = N_{\text{rand}}$. In the latter case, a neuron in the network was randomly chosen and the current was applied in that neuron and its neighbors, which formed a package of neurons that received the application with $|\mathcal{G}| = N_{\text{package}}$. Figure 5 depicts an example of a subset $\mathcal{G}$ with $\lambda_{0,i}/\lambda_0 = 1$, in this case $|\mathcal{G}| = 1000$. The Figure 5a shows a subset of random neurons and Figure 5b a subset of a package of neurons.

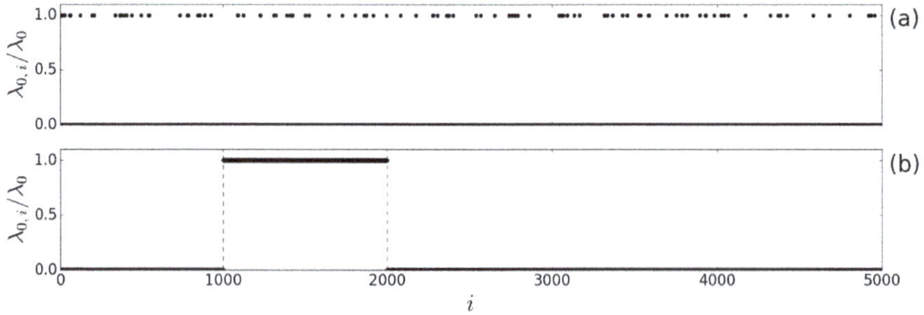

Figure 5. Amplitude of the external current $\lambda_{0,i}$ for: (**a**) random subset of neurons; and (**b**) package subset of neurons.

Figure 6 depicts the mean value of Kuramoto order parameter $\langle R \rangle \times \varepsilon$ as a function of the order of the subset $|\mathcal{G}|$ with $\lambda_0 = 0.1$. In Figure 6a a subset of the neurons with higher connectivity in the network with $|\mathcal{G}| = N_{\text{hubs}}$ is chosen, Figure 6b a subset of random neurons with $|\mathcal{G}| = N_{\text{rand}}$ and Figure 6b a package of neurons with $|\mathcal{G}| = N_{\text{package}}$. Note that the three surfaces have the same shape. In this case, when the order of $|G| \gtrsim 2000$, the PS was suppressed, that is, now the network depicted a monotonic evolution of the synchronization as ε increased.

Another strategy consisted in the application of a delayed mean field signal $\bar{V}$ over the network

$$\xi(t) = \xi_0 \bar{V}(t - t_{\text{delay}}), \tag{22}$$

where ξ_0 (similarly to λ_0) is the current amplitude given by 10^{-4} $\mu\text{A/cm}^2$, and t_{delay} (ms) is the delay time between the generation of the mean field and the re-application of the signal. $\bar{V}$ is the mean field potential of the network, which is defined by

$$\bar{V}(t) = \frac{1}{N} \sum_{i=1}^{N} V_i(t). \tag{23}$$

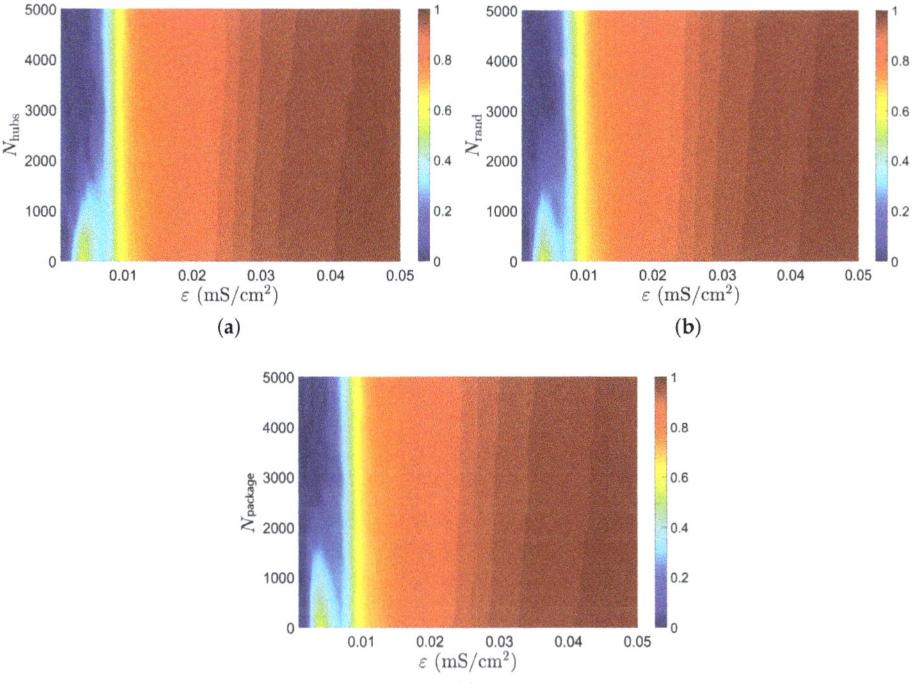

Figure 6. $\langle R \rangle$ as functions of ε and the order of the subset $|\mathcal{G}|$ which is applied an external pulsed current with λ_0 ($i \subset \mathcal{G}$) for three different subsets $\mathcal{G}$: (**a**) $\mathcal{G}$ consists of the neurons with higher connectivity of the network (hubs); (**b**) $\mathcal{G}$ consists of random neurons; and (**c**) $\mathcal{G}$ consists of packages of neurons.

Figure 7 depicts the $\bar{V}(t)$ for three different values of coupling ε. In Figure 7a, $\varepsilon = 0.001$, the mean field display a small amplitude variation since the network presents an unsynchronized behavior. In Figure 7b, $\varepsilon = 0.007$ and, in Figure 7c, $\varepsilon = 0.020$, the mean field display an oscillatory behavior, since in this regime the neurons of the network presents a signal of partial synchronization.

In this approach, $\bar{V} < 0$, $\xi_0 > 0$ ($\xi_0 < 0$) characterizes an inhibitory (excitatory) current which decreases (increases) the membrane potential V_i. The natural period of the Hodgkin–Huxley-type neuron with the parameters in Table 1 is $t_0 \approx 1{,}250$ ms. In Figure 8, we show how the PS varies with the application of $\xi(t)$ in all the neurons as a function of the amplitude ξ_0 and the coupling parameter ε for three different delay time t_{delay}, in panel Figure 8a, $t_{\text{delay}} = 0$, that is, the mean field affects the network instantaneously; when $-15 < \xi_0 < -5$, the network is characterized by a monotonic evolution of the synchronization the anomalous PS is suppressed; when $\xi_0 > -5$, the anomalous regime still occurs;

and, particularly, when $\xi_0 > 5$ the synchronization for coupling $\varepsilon < \varepsilon^*$ is amplified with $\langle R \rangle \approx 0.95$. In Figure 8b, $t_{\text{delay}} = 500$, which is $t_{\text{delay}} \approx t_0/2$, and the mean field with delay $\bar{V}(t - t_0/2)$ is in anti-phase with the $\bar{V}(t)$, for $\xi_0 < -7.5$ the anomalous PS is suppressed, otherwise the network still depicts a non-monotonic evolution of the $\langle R \rangle$ which characterizes abnormal synchronization. In Figure 8c, $t_{\text{delay}} = 1000 \approx t_0$, as expected; the result is similar to Figure 8a because of the oscillatory behavior of the mean field, that is $\bar{V}(t - t_0) \approx \bar{V}(t)$, the numerical value of $\xi(t)$ is the same in both cases.

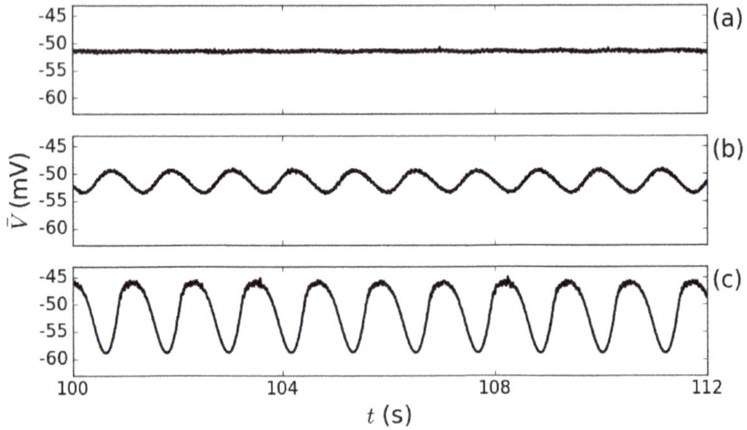

Figure 7. The evolution of the mean field $\bar{V}$ in time for: (**a**) $\varepsilon = 0.001$ (unsynchronized); (**b**) $\varepsilon = 0.007$ ($\approx \varepsilon^*$); and (**c**) $\varepsilon = 0.020$ (synchronized).

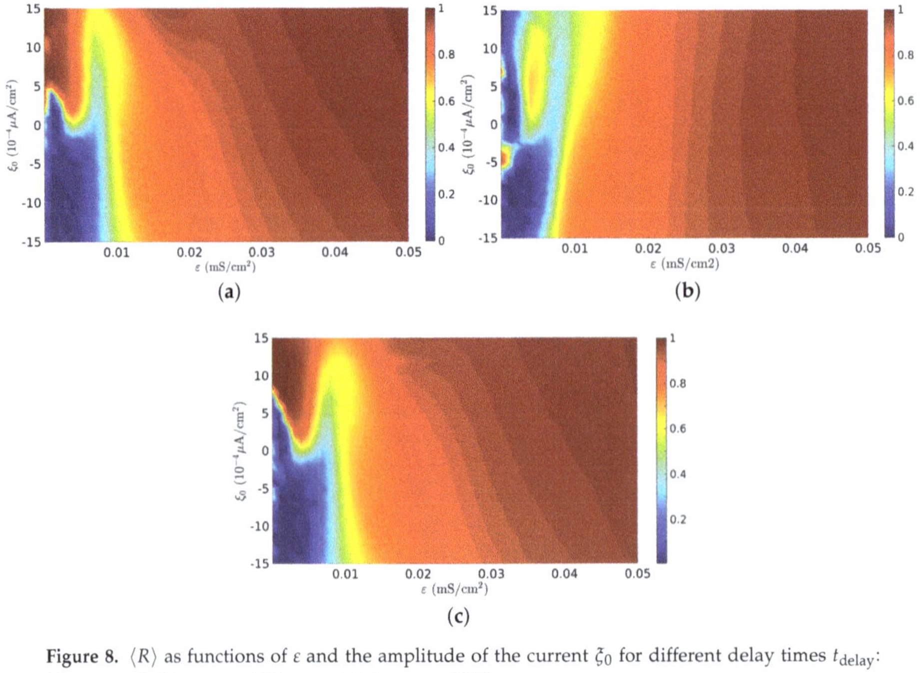

Figure 8. $\langle R \rangle$ as functions of ε and the amplitude of the current ξ_0 for different delay times t_{delay}: (**a**) $t_{\text{delay}} = 0$; (**b**) $t_{\text{delay}} = 500$ ms; and (**c**) $t_{\text{delay}} = 1000$ ms.

5. Conclusions

In this paper, we model a neural network composed of $N = 5000$ Hodgkin–Huxley-type neurons to study the synchronization phenomena. A similar approach have been used to analyze small-world neural networks [31,33,55]. However, the influence of topology plays an important role in the synchronization characteristics [22,35,38]. In this way, we simulated a scale-free network since there are great differences regarding the heterogeneity of the network in comparison to the small-world one [37,38]. It was shown that the scale-free network displays a non-monotonic evolution of the phase synchronization as the coupling between neurons increases. A similar scenario has been observed in small-world networks, which is called "anomalous phase synchronization" [30,31,55], since the traditional behavior should monotonically transition to PS [33]. Especially, Parkinson's disease and some episodes of seizure behavior generated by epilepsy may be associated to anomalous synchronization.

We have proposed two methods of suppression of the anomalous synchronization behavior, both based on treatment for neurological disorders, which consist in the application of an external current in the neurons of the network [9,10].

The first method consists of electrical pulses imposed all over the network. It was shown that, for an amplitude higher than a critical value $\lambda_0 > 0.05$, the anomalous synchronization was suppressed. As a second approach, we studied how the heterogeneity of the scale-free network affects the anomalous PS. We used three different protocols applying the pulsed current in subsets of hubs,; random neurons, and a selected package of neurons. We showed the existence of a threshold, 2000 neurons (40% of the network), which must be disturbed to reach the suppression of anomalous PS in all cases. Such a conclusion implies that the synchronization is related to the individual dynamics of each neuron rather than the network topology [33].

In the second method, only a small fraction $\xi_0 > -0.0005$ μA/cm^2 (with $t_{\text{delay}} = 0$) of the delayed signal of the mean field was applied to all neurons and the abnormal synchronization was suppressed.

Finally, we showed that the delayed signal of the mean field potential had a greater contribution in the region where the suppression was not reached. When $t_{\text{delay}} \approx t_0/2$, it was observed that the anomalous synchronization still persisted but with a lower intensity ($\langle R \rangle \approx 0.6$) compared to the non-delayed scenario ($t_{\text{delay}} = 0$).

Author Contributions: Conceptualization, B.R.R.B., R.C.B., T.L.P. and S.R.L.; Methodology, B.R.R.B., R.C.B., T.L.P. and S.R.L.; Software, B.R.R.B.; Supervision, S.R.L.; Writing—original draft, B.R.R.B., R.C.B., T.L.P. and S.R.L.; and Writing—review and editing, B.R.R.B., R.C.B., T.L.P. and S.R.L.

Acknowledgments: This study was financed in part by the Coordenação de Aperfeiçoamento de Pessoal de Nível Superior—Brasil (CAPES) (Finance Code 001). The authors acknowledge the support of Conselho Nacional de Desenvolvimento Científico e Tecnológico, CNPq, Brazil (grant number 302785/2017-5), Coordenação de Aperfeiçoamento de pessoal de Nível Superior, CAPES (project number 88881.119252/2016-01) and Financiadora de Estudos e Projetos (FINEP).

Conflicts of Interest: The authors declare no conflict of interest.

References

1. Buck, J.; Buck, E. Synchronous fireflies. *Sci. Am.* **1976**, *234*, 74–85.
2. Buck, J. Synchronous rhythmic flashing of fireflies. II. *Q. Rev. Biol.* **1988**, *63*, 265–289. [CrossRef]
3. Mirollo, R.E.; Strogatz, S.H. Synchronization of pulse-coupled biological oscillators. *SIAM J. Appl. Math.* **1990**, *50*, 1645–1662. [CrossRef]
4. Jalife, J. Mutual entrainment and electrical coupling as mechanisms for synchronous firing of rabbit sino-atrial pace-maker cells. *J. Physiol.* **1984**, *356*, 221–243. [CrossRef] [PubMed]
5. Walker, T.J. Acoustic synchrony: two mechanisms in the snowy tree cricket. *Science* **1969**, *166*, 891–894. [CrossRef]
6. Galvan, A.; Wichmann, T. Pathophysiology of parkinsonism. *Clin. Neurophysiol.* **2008**, *119*, 1459–1474. [CrossRef]

7. Hammond, C.; Bergman, H.; Brown, P. Pathological synchronization in Parkinson's disease: Networks, models and treatments. *Trends Neurosci.* **2007**, *30*, 357–364. [CrossRef] [PubMed]
8. Dinstein, I.; Pierce, K.; Eyler, L.; Solso, S.; Malach, R.; Behrmann, M.; Courchesne, E. Disrupted neural synchronization in toddlers with autism. *Neuron* **2011**, *70*, 1218–1225. [CrossRef] [PubMed]
9. Popovych, O.V.; Tass, P.A. Desynchronizing electrical and sensory coordinated reset neuromodulation. *Front. Hum. Neurosci.* **2012**, *6*, 58. [CrossRef]
10. Popovych, O.V.; Tass, P.A. Control of abnormal synchronization in neurological disorders. *Front. Neurol.* **2014**, *5*, 268. [CrossRef] [PubMed]
11. Kringelbach, M.L.; Jenkinson, N.; Owen, S.L.F.; Aziz, T.Z. Translational principles of deep brain stimulation. *Nat. Rev. Neurosci.* **2007**, *8*, 623–635. [CrossRef]
12. Grossman, N.; Bono, D.; Dedic, N.; Kodandaramaiah, S.B.; Rudenko, A.; Suk, H.J.; Cassara, A.M.; Neufeld, E.; Kuster, N.; Tsai, L.H.; et al. Noninvasive deep brain stimulation via temporally interfering electric fields. *Cell* **2017**, *169*, 1029–1041. [CrossRef] [PubMed]
13. McConnell, G.C.; So, R.Q.; Hilliard, J.D.; Lopomo, P.; Grill, W.M. Effective deep brain stimulation suppresses low-frequency network oscillations in the basal ganglia by regularizing neural firing patterns. *J. Neurosci.* **2012**, *32*, 15657–15668. [CrossRef]
14. Weinberger, M.; Mahant, N.; Hutchison, W.D.; Lozano, A.M.; Moro, E.; Hodaie, M.; Lang, A.E.; Dostrovsky, J.O. Beta oscillatory activity in the subthalamic nucleus and its relation to dopaminergic response in Parkinson's disease. *J. Neurophysiol.* **2006**, *96*, 3248–3256. [CrossRef] [PubMed]
15. Benabid, A.L.; Pollak, P.; Gao, D.; Hoffmann, D.; Limousin, P.; Gay, E.; Payen, I.; Benazzouz, A. Chronic electrical stimulation of the ventralis intermedius nucleus of the thalamus as a treatment of movement disorders. *J. Neurosurg.* **1996**, *84*, 203–214. [CrossRef]
16. Kandel, E.R.; Schwartz, J.H.; Jessell, T.M.; Siegelbaum, S.A.; Hudspeth, A.J. *Principles of Neural Science*; McGraw-Hill: New York, NY, USA, 2000; Volume 4.
17. Nicholls, J.G.; Martin, A.R.; Wallace, B.G.; Fuchs, P.A. *From Neuron to Brain*; Sinauer Associates: Sunderland, MA, USA, 2001; Volume 271.
18. Strogatz, S.H. Exploring complex networks. *Nature* **2001**, *410*, 268–276. [CrossRef]
19. Bassett, D.S.; Bullmore, E.D. Small-world brain networks. *Neuroscientist* **2006**, *12*, 512–523. [CrossRef]
20. Zhou, W.; Yang, J.; Zhou, L.; Tong, D. *Stability and Synchronization Control of Stochastic Neural Networks*; Springer: Berlin, Germany, 2015.
21. Eguiluz, V.M.; Chialvo, D.R.; Cecchi, G.A.; Baliki, M.; Apkarian, A.V. Scale-free brain functional networks. *Phys. Rev. Lett.* **2005**, *94*, 018102. [CrossRef]
22. Budzinski, R.; Boaretto, B.; Rossi, K.; Prado, T.; Kurths, J.; Lopes, S. Nonstationary transition to phase synchronization of neural networks induced by the coupling architecture. *Phys. A Stat. Mech. Appl.* **2018**, *507*, 321–334. [CrossRef]
23. Budzinski, R.; Boaretto, B.; Prado, T.; Lopes, S. Phase synchronization and intermittent behavior in healthy and Alzheimer-affected human-brain-based neural network. *Phys. Rev. E* **2019**, *99*, 022402. [CrossRef]
24. Feudel, U.; Neiman, A.; Pei, X.; Wojtenek, W.; Braun, H.A.; Huber, M.; Moss, F. Homoclinic bifurcation in a Hodgkin–Huxley model of thermally sensitive neurons. *Chaos Interdiscip. J. Nonlinear Sci.* **2000**, *10*, 231–239. [CrossRef]
25. Braun, H.A.; Huber, M.T.; Dewald, M.; Schäfer, K.; Voigt, K. Computer simulations of neuronal signal transduction: the role of nonlinear dynamics and noise. *Int. J. Bifurc. Chaos* **1998**, *8*, 881–889. [CrossRef]
26. Braun, H.A.; Dewald, M.; Schäfer, K.; Voigt, K.; Pei, X.; Dolan, K.; Moss, F. Low-dimensional dynamics in sensory biology 2: Facial cold receptors of the rat. *J. Comput. Neurosci.* **1999**, *7*, 17–32. [CrossRef]
27. Hodgkin, A.L.; Huxley, A.F. A quantitative description of membrane current and its application to conduction and excitation in nerve. *J. Physiol.* **1952**, *117*, 500–544. [CrossRef]
28. Shorten, P.R.; Wall, D.J.N. A Hodgkin–Huxley model exhibiting bursting oscillations. *Bull. Math. Biol.* **2000**, *62*, 695–715. [CrossRef]
29. Xu, K.; Maidana, J.P.; Castro, S.; Orio, P. Synchronization transition in neuronal networks composed of chaotic or non-chaotic oscillators. *Sci. Rep.* **2018**, *8*, 8370. [CrossRef]
30. Budzinski, R.C.; Boaretto, B.R.R.; Prado, T.L.; Lopes, S.R. Detection of nonstationary transition to synchronized states of a neural network using recurrence analyses. *Phys. Rev. E* **2017**, *96*, 012320. [CrossRef]

31. Boaretto, B.R.R.; Budzinski, R.C.; Prado, T.L.; Kurths, J.; Lopes, S.R. Suppression of anomalous synchronization and nonstationary behavior of neural network under small-world topology. *Phys. A Stat. Mech. Appl.* **2018**, *497*, 126–138. [CrossRef]
32. Blasius, B.; Montbrió, E.; Kurths, J. Anomalous phase synchronization in populations of nonidentical oscillators. *Phys. Rev. E* **2003**, *67*, 035204. [CrossRef]
33. Boaretto, B.R.R.; Budzinski, R.C.; Prado, T.L.; Kurths, J.; Lopes, S.R. Neuron dynamics variability and anomalous phase synchronization of neural networks. *Chaos Interdiscip. J. Nonlinear Sci.* **2018**, *28*, 106304. [CrossRef]
34. Budzinski, R.C.; Boaretto, B.R.R.; Prado, T.L.; Lopes, S.R. Temperature dependence of phase and spike synchronization of neural networks. *Chaos Solitons Fractals* **2019**, *123*, 35–42. [CrossRef]
35. Gómez-Gardenes, J.; Moreno, Y.; Arenas, A. Paths to synchronization on complex networks. *Phys. Rev. Lett.* **2007**, *98*, 034101. [CrossRef]
36. Zhang, X.; Zou, Y.; Boccaletti, S.; Liu, Z. Explosive synchronization as a process of explosive percolation in dynamical phase space. *Sci. Rep.* **2014**, *4*, 5200. [CrossRef]
37. Newman, M.E.J. Assortative mixing in networks. *Phys. Rev. Lett.* **2002**, *89*, 208701. [CrossRef]
38. Liu, W.; Wu, Y.; Xiao, J.; Zhan, M. Effects of frequency-degree correlation on synchronization transition in scale-free networks. *Eur. Phys. Lett.* **2013**, *101*, 38002. [CrossRef]
39. Schiff, S.J.; Jerger, K.; Duong, D.H.; Chang, T.; Spano, M.L.; Ditto, W.L. Controlling chaos in the brain. *Nature* **1994**, *370*, 615–620. [CrossRef]
40. Schöll, E.; Hiller, G.; Hövel, P.; Dahlem, M.A. Time-delayed feedback in neurosystems. *Philos. Trans. R. Soc. A Math. Phys. Eng. Sci.* **2009**, *367*, 1079–1096. [CrossRef]
41. Batista, C.A.S.; Lopes, S.R.; Viana, R.L.; Batista, A.M. Delayed feedback control of bursting synchronization in a scale-free neuronal network. *Neural Netw.* **2010**, *23*, 114–124. [CrossRef]
42. Rosenblum, M.; Pikovsky, A. Delayed feedback control of collective synchrony: An approach to suppression of pathological brain rhythms. *Phys. Rev. E* **2004**, *70*, 041904. [CrossRef]
43. Kuramoto, Y. *Chemical Oscillations, Waves, and Turbulence*; Springer Science & Business Media: Berlin/Heidelberg, Germany, 2012; Volume 19.
44. Barabási, A.; Albert, R.; Jeong, H. Scale-free characteristics of random networks: The topology of the world-wide web. *Phys. A Stat. Mech. Appl.* **2000**, *281*, 69–77. [CrossRef]
45. Albert, R.; Barabási, A.L. Statistical mechanics of complex networks. *Rev. Mod. Phys.* **2002**, *74*, 47. [CrossRef]
46. Chialvo, D.R. Critical brain networks. *Phys. A Stat. Mech. Appl.* **2004**, *340*, 756–765. [CrossRef]
47. Hagberg, A.; Chult, D.; Swart, P. Exploring network structure, dynamics, and function using NetworkX. In Proceedings of the 7th Python in Science Conference, Pasadena, CA, USA, 19–24 August 2008.
48. Destexhe, A.; Mainen, Z.F.; Sejnowski, T.J. An efficient method for computing synaptic conductances based on a kinetic model of receptor binding. *Neural Comput.* **1994**, *6*, 14–18. [CrossRef]
49. Cohen, S.D.; Hindmarsh, A.C.; Dubois, P.F. CVODE, a stiff/nonstiff ODE solver in C. *Comput. Phys.* **1996**, *10*, 138–143. [CrossRef]
50. Coombes, S.; Bressloff, P.C. *Bursting: The Genesis of Rhythm in the Nervous System*; World Scientific: Singapore, 2005.
51. Prado, T.L.; Lopes, S.R.; Batista, C.A.S.; Kurths, J.; Viana, R.L. Synchronization of bursting Hodgkin–Huxley-type neurons in clustered networks. *Phys. Rev. E* **2014**, *90*, 032818. [CrossRef]
52. Ivanchenko, M.V.; Osipov, G.V.; Shalfeev, V.D.; Kurths, J. Phase synchronization in ensembles of bursting oscillators. *Phys. Rev. Lett.* **2004**, *93*, 134101. [CrossRef]
53. Kuramoto, Y. Collective synchronization of pulse-coupled oscillators and excitable units. *Phys. D Nonlinear Phenom.* **1991**, *50*, 15–30. [CrossRef]
54. Perlmutter, J.S.; Mink, J.W. Deep brain stimulation. *Annu. Rev. Neurosci.* **2006**, *29*, 229–257. [CrossRef]
55. Budzinski, R.C.; Boaretto, B.R.R.; Prado, T.L.; Lopes, S.R. Synchronization domains in two coupled neural networks. *Commun. Nonlinear Sci. Numer. Simul.* **2019**, *75*, 140–151. [CrossRef]

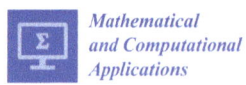

Mathematical and Computational Applications

MDPI

Article

Time Recurrence Analysis of a Near Singular Billiard

Rodrigo Simile Baroni *, Ricardo Egydio de Carvalho, Bruno Castaldi and Bruno Furlanetto

Institute of Geosciences and Exact Sciences (IGCE), São Paulo State University (UNESP), Av. 24A 1515, Bela Vista, Rio Claro-SP, Brazil; ricardo.egydio@unesp.br (R.E.d.C.); brunocastaldi@gmail.com (B.C.); bruno.furlanetto@unesp.br (B.F.)

* Correspondence: r.baroni@unesp.br

Received: 11 March 2019; Accepted: 7 May 2019; Published: 8 May 2019

Abstract: Billiards exhibit rich dynamical behavior, typical of Hamiltonian systems. In the present study, we investigate the classical dynamics of particles in the eccentric annular billiard, which has a mixed phase space, in the limit that the scatterer is point-like. We call this configuration the near singular, in which a single initial condition (IC) densely fills the phase space with straight lines. To characterize the orbits, two techniques were applied: (i) Finite-time Lyapunov exponent (FTLE) and (ii) time recurrence. The largest Lyapunov exponent λ was calculated using the FTLE method, which for conservative systems, $\lambda > 0$ indicates chaotic behavior and $\lambda = 0$ indicates regularity. The recurrence of orbits in the phase space was investigated through recurrence plots. Chaotic orbits show many different return times and, according to Slater's theorem, quasi-periodic orbits have at most three different return times, the bigger one being the sum of the other two. We show that during the transition to the near singular limit, a typical orbit in the billiard exhibits a sharp drop in the value of λ, suggesting some change in the dynamical behavior of the system. Many different recurrence times are observed in the near singular limit, also indicating that the orbit is chaotic. The patterns in the recurrence plot reveal that this chaotic orbit is composed of quasi-periodic segments. We also conclude that reducing the magnitude of the nonlinear part of the system did not prevent chaotic behavior.

Keywords: recurrence time; Slater's theorem; Lyapunov exponent; point scatterer; annular billiard

1. Introduction

The problem of a one-point particle, or an ensemble of non-interacting point particles, moving with constant velocity inside a bounded region, subject to elastic collisions with the boundaries, is generically known as a billiard. Such dynamical systems are described by nonlinear mappings and have great interest in several branches of physics [1–6]. Even billiards with simple geometry exhibit rich dynamical behaviors, typical of Hamiltonian systems, and depending on the geometry of the boundaries and the control parameters, the associated phase space can be: I—regular, consisting of quasi-periodic or periodic orbits lying on Kolmogorov–Arnold–Moser (KAM) tori; II—chaotic, with orbits that densely fill the whole phase space; or III—mixed, in which regular motion coexists with chaotic motion. This is the case with most non-integrable Hamiltonian systems, wherein the characterization of orbits as regular or chaotic is of great interest [7].

As a consequence of mixed phase space, some chaotic orbits that come sufficiently close to a KAM island tend to spend a long time around that region, almost behaving as regular orbits. Following this transient time, these orbits return to the chaotic sea. This phenomenon is known as stickiness and it influences the transport properties of the system [6,8–10].

To distinguish chaotic orbits from regular ones, a standard approach that has been frequently employed is the estimation of Lyapunov exponents [6,11]. For conservative systems, regular orbits have null Lyapunov exponents, while chaotic orbits have at least one positive Lyapunov exponent. For

Hamiltonian systems, the sum of the exponents must be zero in order to satisfy Liouville's theorem. In numerical applications, finite-time Lyapunov exponents (FTLEs) [6,11–13] can be calculated, and although they are suitable to investigate systems with a mixed phase space, the existence of stable islands delays the convergence of chaotic orbits, wherein simulations of high computational effort are required to calculate the exponents. The distribution of the FTLEs over initial conditions carries information about the system dynamics. For completely chaotic systems, a Gaussian distribution is expected. If a small amount of regular structures are present in the phase space, the orbits affected by stickiness will have smaller FTLE values, which induces a tail in the distribution, making it asymmetric. If a large amount of regular structures are present, the system has strong sticky motion and the distribution of the FTLEs is multimodal [14,15].

In the late 1960s, an interesting and not well-known theorem was proposed by Slater [16], whose nonlinear version was well adapted in [6,17]. It states that it is possible to distinguish quasi-periodic dynamics from chaotic dynamics by analyzing the Poincaré return times, also called recurrence times, and that for any connected interval of size ε, there are no more than three different return times, the largest of them being the sum of the other two.

The technique of recurrence plots (RPs) [18,19] has been applied for the visualization and analysis of nonlinear experimental data in many different fields, from biological sciences [20–22] to complex systems [23–25]. This tool was first introduced in the context of dynamical systems to visualize the recurrence of trajectories in the phase space [19]. As chaotic and quasi-periodic orbits have different recurrence properties, RPs allow us to distinguish the behavior of these orbits, identify the interval of time that a chaotic orbit has been trapped by stickiness, and even quantify stickiness by using recurrence quantification analysis [26].

Quantum mechanical versions of billiards provide useful models to investigate the quantum properties of certain systems, as it is possible to identify signatures of the structures present in the phase space of their classical analogue in the quantum wave functions [27–29]. The distribution of spacings between the neighbor energy levels of a quantum billiard obeys a Poisson distribution for classically integrable systems and a Wigner distribution (or GOE distribution) for classically chaotic systems. In the early 1990s, a billiard with a point-like scatterer was first proposed to investigate the quantum aspects of systems whose corresponding classical limit was between integrable and chaotic. This class of systems is known as singular (or Šeba) billiards [27] and the distance between its neighbor energy levels obeys a Poisson distribution.

A zero-size scatterer is an idealized limit of a small obstacle [29]. In this work, we investigate the classical dynamics of a billiard with a very small scatterer through its recurrence properties and estimations of the largest Lyapunov exponent (LLE). The chosen model is the annular billiard, which has been the subject of many analytical and numerical studies [28,30–35]. It consists of a particle confined in the region between two circumscribed circumferences of radii R and r, with $r < R$. When the circumferences are concentric, the energy and angular momentum are both conserved and therefore the system is integrable, since the plane billiard has two degrees of freedom. If the circumferences are eccentric, the angular momentum is altered with the collisions with the inner circle and chaotic behavior may be observed. The limit of the very small scatterer is obtained by choosing $r \ll R$ and we have called this limit to be a near-singular billiard.

It has been postulated in [36] that it is not possible to remove the chaotic behavior of chaotic nonlinear systems by reducing the magnitude of the nonlinear part. The author has arrived at this conclusion by examining a few cases of chaotic systems, described by one-dimensional mappings, when the nonlinear part is reduced. This work also aims to verify the validity of this proposition for our model, which is described by a two-dimensional nonlinear mapping.

2. Materials and Methods

In this section we will discuss the annular billiard model, the numerical method applied to estimate the largest Lyapunov exponent and the technique of RPs.

The system of interest consists of a classical particle confined in the annular region limited by two circumscribed circumferences. The geometry may be concentric or eccentric. The radius of the outer circle is defined as $R = 1$, the radius of the inner circle (the scatterer) as r, and the eccentricity as d. See Figure 1 for an example of the annular billiard geometry.

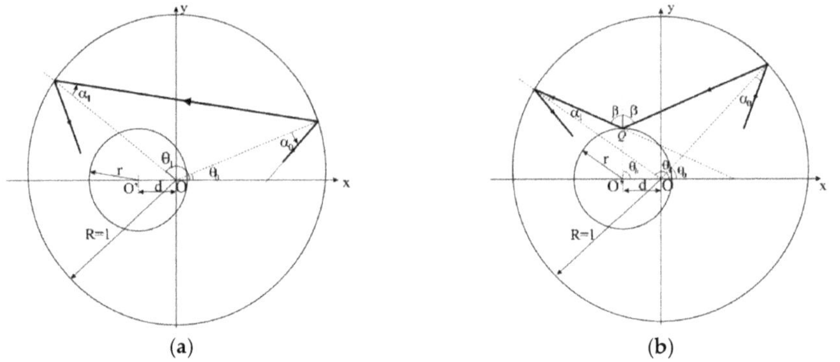

Figure 1. Geometric scheme of a particle in the annular billiard. (**a**) Type A movement, where the particle does not collide with the scatterer between two successive collisions with the outer boundary. (**b**) Type B movement, in which a collision with the scatterer occurs (adapted from [31]).

On the billiard, a particle moves freely in a straight line with constant velocity until it elastically collides with a boundary. Succeeding a collision, the particle´s trajectory is specularly reflected, i.e., the angle of reflection is equal to the angle of incidence. From Figure 1, we can see that the position at a collision with the outer boundary is determined by the angle of incidence α and the arc-angle θ, for $\theta \in [-\pi, \pi]$ and $\alpha \in [-\pi/2, \pi/2]$. The map that describes the dynamics projects the coordinates (θ_n, α_n) of a collision with the outer boundary in the coordinates $(\theta_{n+1}, \alpha_{n+1})$ of the next collision.

There are two kinds of motion which are distinguished by the tangency condition, which reads as follows:

$$\left|\sin(\alpha_n) - d\sin(\theta_n - \alpha_n)\right| \leq r. \tag{1}$$

If the combination of (θ_n, α_n) does not satisfy Condition (1), the dynamics is described by the map M_A and therefore the movement is of type A.

Type A: Between two successive collisions with the outer boundary, the particle does not collide with the scatterer, and so the map M_A reads as follows:

$$M_A : \begin{cases} \alpha_{n+1} = \alpha_n \\ \theta_{n+1} = \pi + \theta_n - 2\alpha_n \end{cases}. \tag{2}$$

On the other hand, if Condition (1) is satisfied, the dynamics is described by the map M_B and therefore the movement is of type B.

Type B: Between two successive collisions with the external circle, the particle collides with the internal circle, and the dynamics is given by

$$M_B : \begin{cases} \alpha_{n+1} = \sin^{-1}(r\sin\beta - d\sin\theta_a) \\ \theta_{n+1} = -\alpha_{n+1} + \theta_a \end{cases}, \tag{3}$$

where

$$\begin{aligned} \theta_a &= 2\beta + \theta_n - \alpha_n \\ \beta &= \sin^{-1}\left\{\frac{1}{r}[\sin(\alpha_n) - d\sin(\theta_n - \alpha_n)]\right\}. \end{aligned} \tag{4}$$

The mapping equations have been obtained via geometrical considerations and a more detailed discussion can be found in [28].

The trajectories that will never cross the caustic (an auxiliary circle centered at the origin, with a radius of $r + d$) or never hit the scatterer are the so-called whispering gallery orbits (WGO). In order to keep the phase space area filled by the WGO constant, when varying the eccentricity, the radius of the caustic in the simulations was fixed as $r + d = 0.7005$. The canonically conjugate pair of variables used to determine the Poincaré section were $S = \sin\alpha$, with $|S| \leq 1$, and $L = (\theta/2\pi)$, with $|L| \leq (1/2)$, collected at every collision with the outer boundary. It is worth emphasizing that S is the angular momentum of the particle with respect to the origin of the outer boundary at the moment of collision.

The conditions to generate periodic orbits in the concentric annular billiard are well-known. For trajectories that do not hit the scatterer, the dynamical behavior is the same as in the circular billiard, and the initial condition (θ_0, α_0) generates a periodic orbit if

$$\alpha_0 = \left(\frac{1}{2} - \frac{K}{N}\right)\pi, \tag{5}$$

where N is the period of the orbit and K is the number of turns made by the particle around the billiard. Due to the billiard's symmetry, any value of θ_0 generates an orbit with the same period.

For trajectories that hit the inner boundary, the periodic orbit condition is

$$\alpha_0 = \arctan\left[\frac{\sin(\pi/N)}{1/r - \cos(\pi/N)}\right]. \tag{6}$$

The Lyapunov exponents quantify the average expansion or contraction of a small volume of initial conditions. Given a dynamical system represented by the orbit $\{\vec{v_i}\}$, with $i = 0, \ldots, n$ being the number of the iteration, in a phase space of arbitrary dimension, the stability of the orbit may be verified through the evolution of a satellite orbit named $\{\vec{v_i}'\}$. Initially, both orbits are separated by an infinitesimal distance δv_0, with $\delta v_0 = ||\vec{v_0} - \vec{v_0}'||$, where the zero index indicates the initial iteration. For chaotic orbits, this distance grows exponentially with time, according to

$$\delta v_n \propto \delta v_0 e^{\lambda n}, \tag{7}$$

where λ is the exponential expansion rate. This allows us to define the Lyapunov exponent as

$$\lambda = \lim_{n\to\infty} \lim_{\delta v_0 \to 0} \frac{1}{n} \ln\left(\frac{\delta v_n}{\delta v_0}\right). \tag{8}$$

This equation estimates the average exponential separation rate between the original orbit and its satellite one. If λ is greater than zero, the orbit is chaotic. Otherwise, it is periodic or quasi-periodic.

The numerical procedure to estimate the largest Lyapunov exponent in the annular billiard with the FTLE method can be described as follows: First, we truncate the time evolution at a finite, but long, time. Then, given an initial orbit (L_0, S_0) and a satellite orbit (L_0', S_0'), initially separated by the distance δ_0, set as $\delta_0 = 10^{-10}$, we evolve both orbits according to the mapping dynamics. After a defined rescaling time ℓ, we measure the distance between them as

$$\delta_\ell = \sqrt{(\Delta L)^2 + (\Delta S)^2}, \tag{9}$$

where $\Delta L = L_\ell - L_\ell'$ and $\Delta S = S_\ell - S_\ell'$. The distance between orbits δ_ℓ is used to rescale the satellite orbit in the same direction of the original orbit and then the procedure is restarted with

$$\begin{aligned} L_0' &= L_\ell + \left[\delta_0(L_\ell' - L_\ell)\right]/\delta_\ell \\ S_0' &= S_\ell + \left[\delta_0(S_\ell' - S_\ell)\right]/\delta_\ell. \end{aligned} \tag{10}$$

The FTLE is computed considering k's successive ℓ time intervals, until the predetermined finite time is reached, with the expression

$$\lambda = \frac{1}{k\ell}\sum_{j=1}^{k} \ln\left(\frac{\delta_j}{\delta_0}\right). \tag{11}$$

We will now discuss in a general way the technique of RPs. Once again, given the orbit $\{\vec{v_i}\}$, with $i = 0, \ldots, n$ being the number of the iteration, we can compute the recurrence matrix,

$$\mathbf{R}_{i,j} = \Theta(\varepsilon - ||\vec{v_i} - \vec{v_j}||), \tag{12}$$

where ε is a predefined threshold, $\Theta(\cdot)$ is the Heaviside function, and $||\cdot||$ is a norm defining the distance between two points, such that $\mathbf{R}_{i,j} = 0$ if $\varepsilon - ||\vec{v_i} - \vec{v_j}|| < 0$ and $\mathbf{R}_{i,j} = 1$ if $\varepsilon - ||\vec{v_i} - \vec{v_j}|| > 0$. $\mathbf{R}_{i,j}$ is a binary square matrix, with elements of 'zeros' and 'ones', and its graphical representation is called a 'recurrence plot'. The value 'one' is encoded by a black point, i.e., the distance between the respective phase space points is smaller than ε, which means that at the i-th iteration the orbit returned to the region of the phase space where it was at the j-th iteration, after $i - j$ iterations. The value of 'zero' is encoded by a white point, i.e., the distance between the phase space points at the j-th and i-th iteration is larger than ε, and therefore the orbit had not yet returned to the previously visited reference region of the phase space. A white vertical line between two points provides the recurrence time or Poincaré return time (i.e., the time that takes for the orbit to return to a previous state).

To illustrate the procedure, let us discuss the computation of the first column of the matrix $\mathbf{R}_{i,j}$, i.e., consider $j = 0$, so that the returns to the neighborhood of the phase space point $\vec{v_0}$ will be evaluated. The $n + 1$ elements of the first column of the matrix are given by $\mathbf{R}_{i,0} = \Theta(\varepsilon - ||\vec{v_i} - \vec{v_0}||)$, with $i = 0, \ldots, n$. The first one is $\mathbf{R}_{0,0} = \Theta(\varepsilon - ||\vec{v_0} - \vec{v_0}||) = \Theta(\varepsilon) = 1$, generating a black point in the RP. Let us say that for $i = 1, 2, 3, 4$ the orbit has not returned to the neighborhood of $\vec{v_0}$, which means that the distance between the points $\vec{v_i}$ and $\vec{v_0}$ is greater than ε, and so $\mathbf{R}_{i,0} = \Theta(\varepsilon - ||\vec{v_i} - \vec{v_0}||) = 0$, generating a white point in the RP. Assuming that at the fifth iteration the orbit has returned to the neighborhood of $\vec{v_0}$, the distance between $\vec{v_5}$ and $\vec{v_0}$ is smaller than ε, so $\mathbf{R}_{5,0} = \Theta(\varepsilon - ||\vec{v_5} - \vec{v_0}||) = 1$ and a black point is shown in the RP. The number of iterations that are needed for the orbit to return to the neighborhood of the reference point $\vec{v_0}$ was 5, which is equivalent to the distance between the two consecutive black points in the first column of the RP. The procedure continues until all the elements of the first column of the recurrence matrix are computed, which is then is repeated for all the other columns.

In an RP, a white vertical line between two points provides the recurrence time or Poincaré return time (i.e., the time that it takes for the orbit to return to a previous state). A periodic orbit of period T generates an RP with parallel diagonal lines, all of which are separated by the vertical distance T, as the recurrence occurs at a fixed time interval (Figure 2). On the other hand, a quasiperiodic orbit provides an RP with parallel diagonal lines, with different distances between them. These distances can be, at most, three different ones, and the largest must be the sum of the other two, in accordance with Slater's theorem [16]. RPs of chaotic orbits are quite different however: They show a large number of patterns formed by short diagonal lines and dashed lines and many different return times can be found.

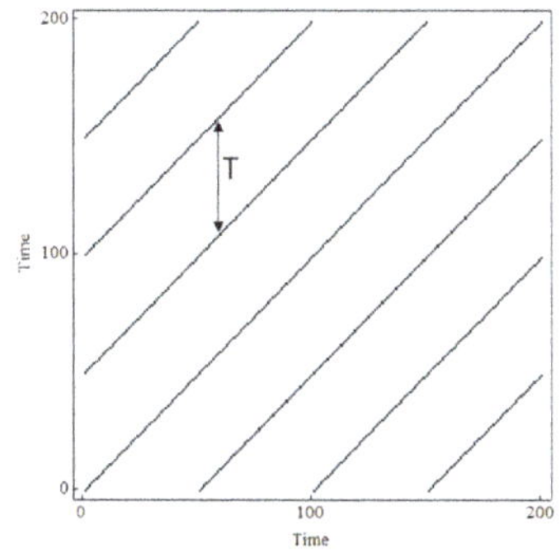

Figure 2. Recurrence plot (RP) of a periodic orbit of period T. This is a graphical representation of the (binary) recurrence matrix **R**, where the value 'one' is encoded by a black dot and the value 'zero' is encoded by a white dot. In the graph, time refers to the countable parameter that represents the evolution of an orbit of the given dynamical system, and the line with arrowheads on both sides denotes the white vertical line of length T, which is equivalent to a recurrence time.

3. Results

This section is organized as follows. In Section 3.1, we present the phase space of the annular billiard for an ensemble of initial conditions, illustrating the role of the eccentricity d as a control parameter that introduces chaos into the system. We also present plots of a single orbit to illustrate the dynamics of the singular billiard in the limit $r \ll R$. In Section 3.2, we present the estimations of the largest Lyapunov exponent for some initial conditions for increasing eccentricity d and, therefore, decreasing inner radius r. In this section we present the distribution of the LLEs in the near singular limit as well. Finally, in Section 3.3, we present RPs of periodic, quasi-periodic and chaotic orbits in the annular billiard and of a typical orbit in the near singular limit. We also discuss the question of sensitive dependence on initial conditions in this limit.

3.1. Phase Space

All simulations presented in this section were carried out with $r + d = 0.7005$ in order to keep the regions occupied by the WGO in the Poincaré section constant, i.e., the regions $|S| \geq 0.7005$. These regions correspond to straight lines and are omitted in the plots.

The set of plots in Figure 3 shows the phase space of the annular billiard for different values of eccentricity d. When the boundaries are concentric, i.e., $d = 0$, the angular momentum, with respect to the origin of the outer boundary, is preserved, the system is integrable, and the phase space is filled with straight lines parallel to the L-axis. In the eccentric case, i.e., $d > 0$, the system is no longer integrable once the angular momentum is no longer a constant of motion, and consequently, structures of resonances and chaos arise. By increasing the value of d the region occupied by the chaotic sea also increases.

In order to illustrate the dynamics in the singular billiard limit, we will now look at only one orbit in the phase space. Figure 4 shows the plots of a single orbit, chosen to be in the chaotic sea for the most values of eccentricity, for different values of increasing eccentricity d and, therefore, decreasing inner radius r. In Figure 4a, we have $d = 0$, so the system is integrable and the orbit fills a straight line. In Figure 4b,c, the orbit is chaotic and densely fills the accessible region of the phase space, where the white regions correspond to KAM islands. In Figure 4d, we have $r = 5 \times 10^{-4}$, so that the radius of the scatterer is four orders of magnitude smaller than the radius of the outer boundary $R = 1$.

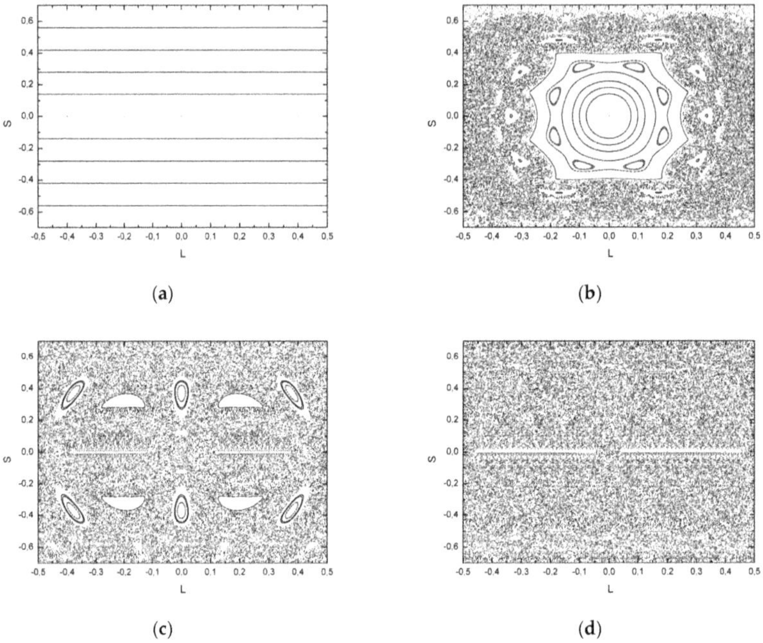

Figure 3. Phase space of the annular billiard with $r + d = 0.7005$. The number of iterations used was $n = 500$. $L = (\theta/2\pi)$ and $S = \sin\alpha$ are the action-angle variables used to determine the Poincaré section. (**a**) $d = 0.00$; (**b**) $d = 0.15$; (**c**) $d = 0.41$; (**d**) $d = 0.55$.

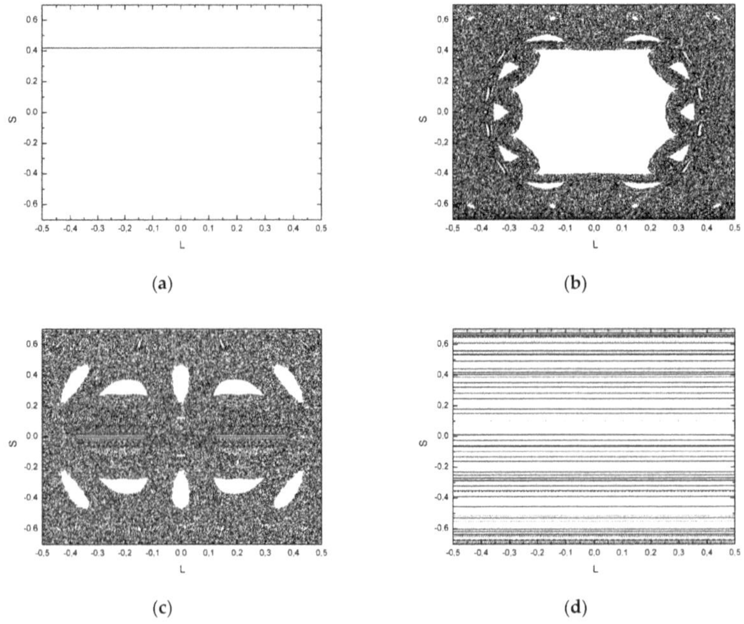

Figure 4. Plots of a single initial condition $(L_0, S_0) = (0.41, 0.42)$ in the phase space of the annular billiard and number of iterations $n = 10^5$. (**a**) $d = 0.00$, $r = 0.7005$; (**b**) $d = 0.15$, $r = 0.5505$; (**c**) $d = 0.41$, $r = 0.2905$; (**d**) $d = 0.70$, $r = 5 \times 10^{-4}$.

In the near singular limit, a single initial condition fills the phase space with straight lines. This happens because for a given initial condition (L_0, S_0) the angular momentum of the particle remains constant and equal to S_0, filling the corresponding straight line (or torus) in the phase space until a collision with the scatterer takes place. When this happens, the angular momentum of the particle is changed, as governed by Equation (3), and another line in the phase space starts to be filled, i.e., another torus is visited. However, once the scatterer is so small, these collisions are rare and the particle spends a long time in the same torus. This process goes on, and by succeeding a sufficiently large number of iterations, the single initial condition densely fills the phase space with straight lines (Figure 4d).

3.2. *Lyapunov Exponent*

The largest Lyapunov exponent λ was calculated with the FTLE method. The finite time chosen was $n = 10^5$, the interval between rescaling $\ell = 1000$, and the initial distance between the original and the satellite orbits $\delta_0 = 10^{-10}$.

The set of plots in Figure 5 exhibits values of λ as function of the eccentricity d and, therefore, the radius of the scatterer r. For $d = 0.7$ we have $r = 5 \times 10^{-4}$, corresponding to the near singular limit.

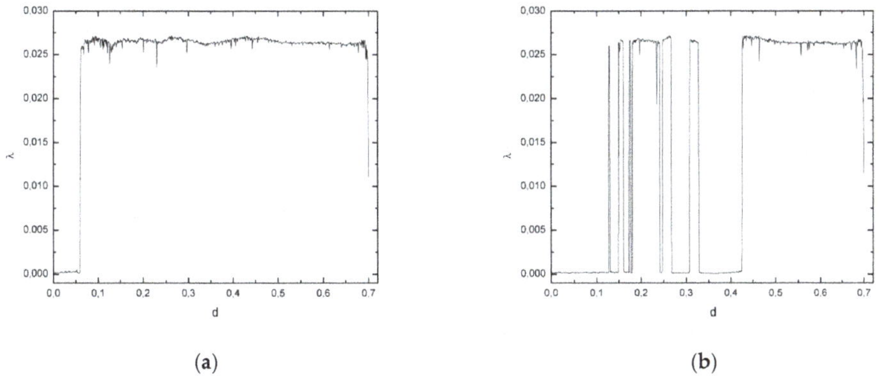

Figure 5. Largest value of Lyapunov exponent versus eccentricity. As the value of d increases, the radius of the inner circle r decreases so that $r + d = 0.7005$. (**a**) $(L_0, S_0) = (0.41,\ 0.42)$; (**b**) $(L_0, S_0) = (0.20,\ 0.30)$.

In Figure 5a, the chosen initial condition was the same as what was used for the plots of Figure 4, which densely fills the phase space with straight lines in the near singular limit. Initially we have $\lambda = 0$, and as the eccentricity increases and chaos begins to appear in the system, the Lyapunov exponent becomes positive. The oscillations observed in λ are presumably due to stickiness, as KAM islands are created and destroyed as d changes, influencing the behavior of the chaotic orbits. In the near singular limit, λ sharply drops for the initial conditions (ICs) in both panels of Figure 5, evidencing a change in the dynamics of the system. Different initial conditions have been used and this scenario is representative of the near singular case.

In Figure 5b, we have the IC $(L_0, S_0) = (0.20,\ 0.30)$, which is not always located in the chaotic sea as the parameter d is varied. There are values of d that give $\lambda = 0$ and this is because the IC is placed at a KAM island for those configurations. Figure 6 exhibits the phase space for two values of eccentricity, where $d = 0.2$ in panel (a), for which both ICs of Figure 5 fall in the chaotic sea and the Lyapunov exponent is positive. In panel (b), $d = 0.378$, for which the Lyapunov exponent calculated for the IC of Figure 5a is positive and for the IC of Figure 5b, it drops to zero. In both images, the green dot corresponds to the IC of Figure 5a and the red dot corresponds to the IC of Figure 5b.

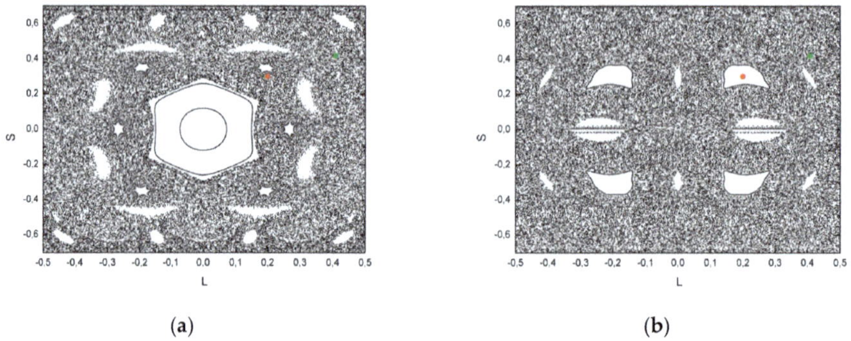

Figure 6. Phase space of the annular billiard with $r + d = 0.7005$. The number of iterations used is $n = 1000$. $L = (\theta/2\pi)$ and $S = \sin\alpha$ are the action-angle variables used to determine the Poincaré section. The red dot corresponds to $(L, S) = (0.20,\ 0.30)$ and the green dot corresponds to $(L, S) = (0.41,\ 0.42)$. (**a**) $d = 0.2$; (**b**) $d = 0.378$.

In Figure 7 we have the distribution of the Lyapunov exponent for 10^6 initial conditions, in the near singular limit, distributed over the (L, S) plane region limited by $|S| \leq 0.7005$ and $|L| \leq (1/2)$. The Gaussian-like distribution with a tail to the left indicates chaotic dynamics with stickiness. The peak of the distribution is close to the values of λ obtained for the ICs of Figure 5 in the near singular limit, confirming that these ICs are representative of the dynamics in this limit.

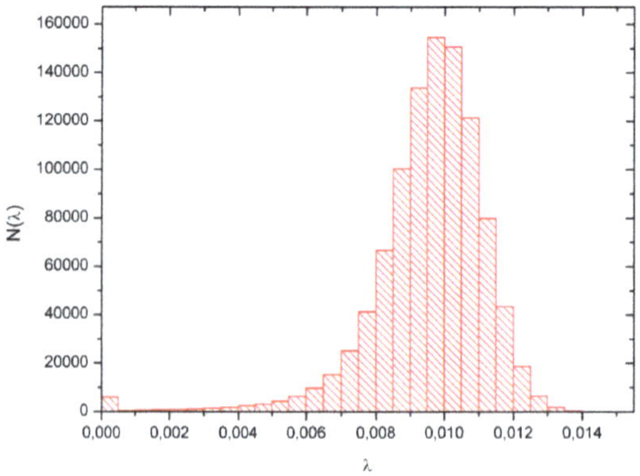

Figure 7. Distribution of the largest value of the Lyapunov exponent over the ICs in the near singular limit ($d = 0.7$). Here, 10^6 ICs were considered. $N(\lambda)$ stands for the number of times that the value λ was calculated.

3.3. Recurrence Plots

Each RP presented here was obtained with different values of the threshold ε, which is the size of the recurrence region. This size has to be evaluated in order to be sure that there are not too few or too many lines in the plots. Although different values of ε provide different return times for a same orbit, Slater's theorem is always obeyed in the case of quasi-periodicity.

In Figure 8a, we have the RP of a periodic orbit of period 5 in the concentric annular billiard, obtained with Equation (6), and in Figure 8b, the same orbit in the phase space. The vertical distance between the lines in the RP of Figure 8a is always the same and exactly 5, i.e., after five collisions with the outer boundary, the orbit will always return to a previously visited reference state. Figure 8b shows the same orbit in the phase space, given in the (L, S) plane. For the concentric configuration, the angular momentum S of the particle is preserved, so that the orbit remains on the same line (torus) of the (L, S) plane. Given the initial condition $(L(n = 0), S(n = 0)) = (L_0, S_0)$, the mapping equations always provide the same five points, as $L(n) = L(n + 5)$.

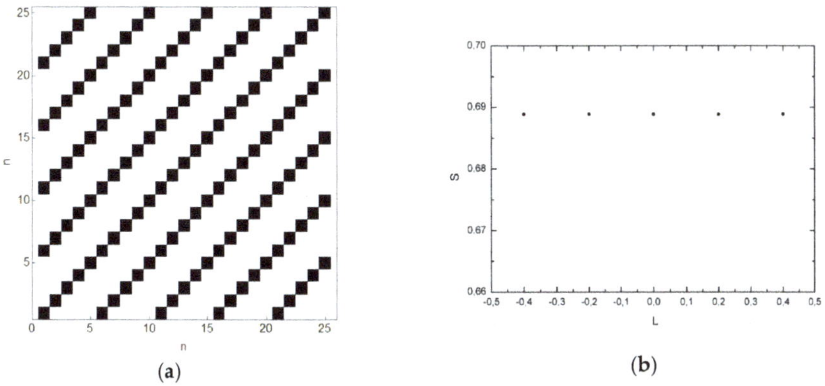

Figure 8. Periodic orbit of period 5 in the concentric annular billiard, the number of iterations is $n = 25$. $(L_0, S_0) = (0.00,\ 0.688858633)$. (**a**) Recurrence plot, $\varepsilon = 0.1$. The distance between two consecutive black points in a column is always 5. (**b**) The corresponding orbit in the phase space. At every 5 iterations, it is back to (L_0, S_0).

Figure 9a shows an RP of a quasi-periodic orbit and Figure 9b shows the respective orbit in the phase space. In contrast to the RP of the periodic orbit presented in Figure 8a, the distances between the diagonal lines of Figure 9a are not constant, revealing different recurrence times for this orbit.

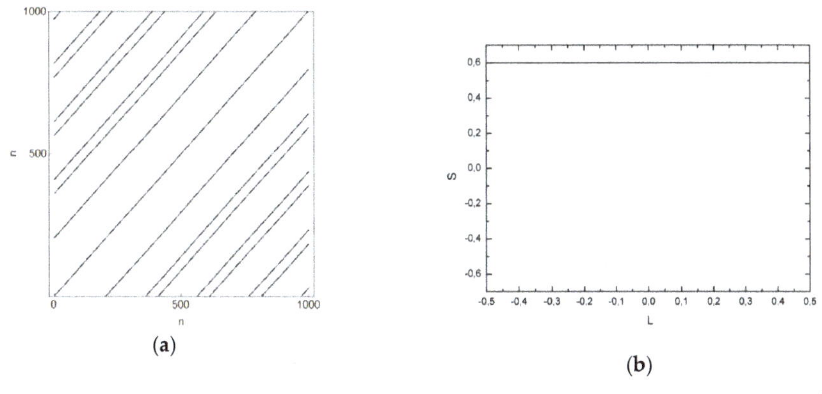

Figure 9. Quasi-periodic orbit for $d = 0.00$, $(L_0, S_0) = (0.0,\ 0.6)$, and number of iterations $n = 10^3$. (**a**) Recurrence plot, $\varepsilon = 0.004$. (**b**) The corresponding orbit in the phase space, filling a straight line.

To visualize the recurrence times implicitly shown in the RP, the following numerical procedure was applied to obtain a histogram of vertical white lines. From the recurrence matrix $R_{i,j}$, we first choose $j = 1$, taking the first column of the matrix. Then, we identify the first and second non-zero elements of that column, $R_{k,1}$ and $R_{l,1}$, and compute the return time $l - k$. The next non-zero element, $R_{m,1}$, is found and another return time, $k - m$, is computed. This process goes on until all the non-zero elements of the first column are considered. The procedure is then repeated for all columns of the matrix. The result can be plotted in a 2D histogram that shows the possible recurrence times as peaks in their respective frequencies of observation.

The histogram in Figure 10 shows three peaks that correspond to return times equal to 49, 155 and 204 iterations, and, in accordance with Slater's theorem, $49 + 155 = 204$.

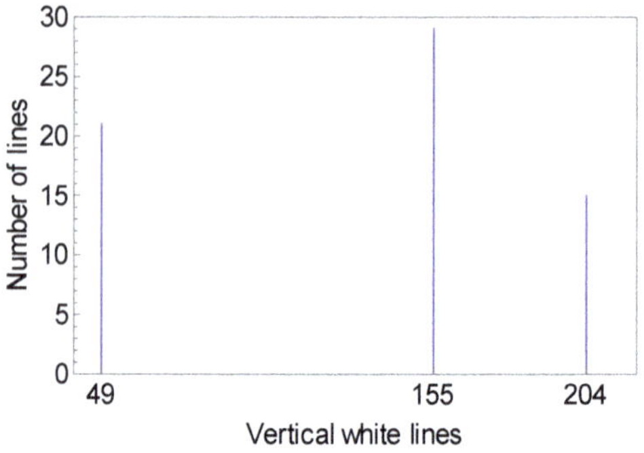

Figure 10. Histogram of vertical white lines in the RP of the quasi-periodic orbit of Figure 9. A vertical white line stands for a computed recurrence time, and the number of lines indicates how many times this recurrence time was observed.

A chaotic orbit of the annular billiard is analyzed in Figure 10. Differently from the periodic and quasi-periodic cases, the RP of the chaotic orbit (Figure 11a) shows complex patterns rather than parallel lines. It is possible to observe structures in the diagonal of the RP, as the one found for $112 \leq n \leq 154$. These structures correspond to times when the orbit was trapped by stickiness. Figure 11b shows the respective orbit in the phase space, where the red dots correspond to the iterations from $n = 112$ to $n = 154$, when the orbit was suffering stickiness.

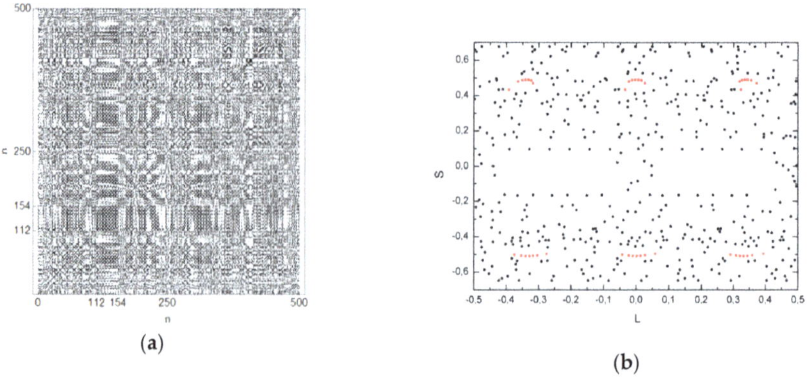

Figure 11. Chaotic orbit in the annular billiard, $d = 0.5$, $(L_0, S_0) = (0.3,\ 0.3)$, and number of iterations $n = 500$. (**a**) Recurrence plot, $\varepsilon = 0.1$, the structure between $n = 112$ and $n = 154$ suggests that the particle was trapped by stickiness phenomena. (**b**) Chaotic orbit filling the phase space. Red dots correspond to iterations that the particle was trapped by stickiness.

The histogram of vertical white lines in Figure 12 shows that a chaotic orbit has many different return times.

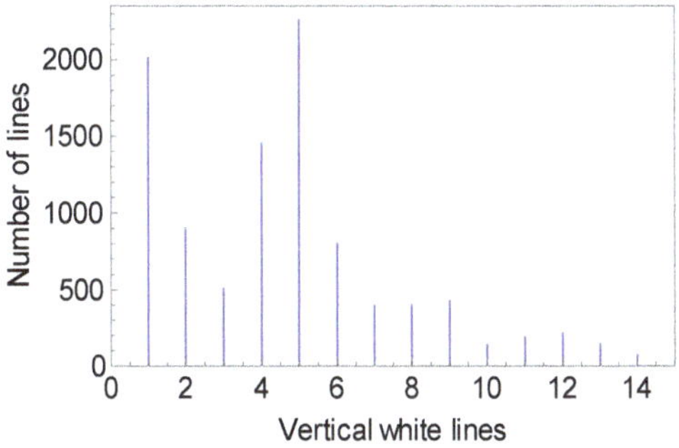

Figure 12. Histogram of vertical white lines in the RP of the chaotic orbit of Figure 11. A vertical white line denotes a computed recurrence time and the number of lines indicates how many times this recurrence time was observed.

We now analyze the RP of an orbit in the near singular billiard. Figure 13 shows the RP corresponding to the orbit of Figure 4d, which densely fills the phase space with straight lines and the corresponding histogram of vertical white lines. The fact that many different return times are observed states that the orbit is chaotic.

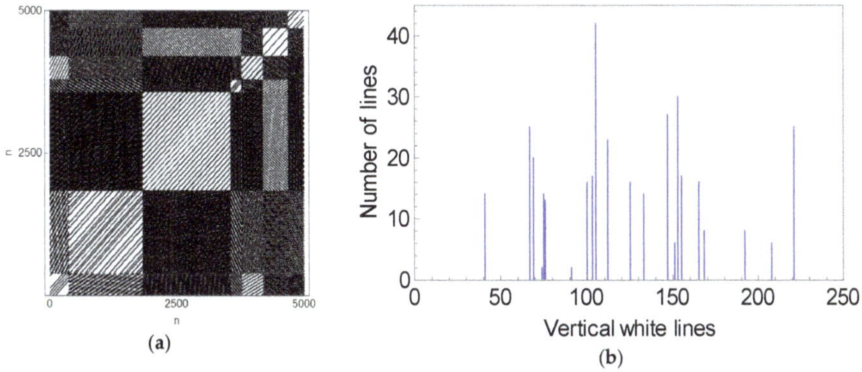

Figure 13. Typical orbit in the near singular billiard, $(L_0, S_0) = (0.41, 0.42)$ and number of iterations $n = 5 \times 10^3$. (**a**) Recurrence plot, $\varepsilon = 0.006$, showing a junction of many quasi-periodic orbits; (**b**) Corresponding histogram of vertical white lines in the RP, a vertical white line stands for a computed recurrence time, and the number of lines indicates how many times this recurrence time was observed.

The RP in Figure 13a is notably different from that of the chaotic orbit in Figure 11a. The orbit in the near singular billiard generates an RP with block-like structures, of straight lines, in a diagonal direction. Separating this RP according to each block (Figure 14a,c), three return times are observed for each of them and they satisfy Slater's theorem. It is important to notice that the number of diagonal lines in the RP depends only on the chosen value of ε, which is why the RPs of Figure 14a,c look different than the respective block-like structures of Figure 13a.

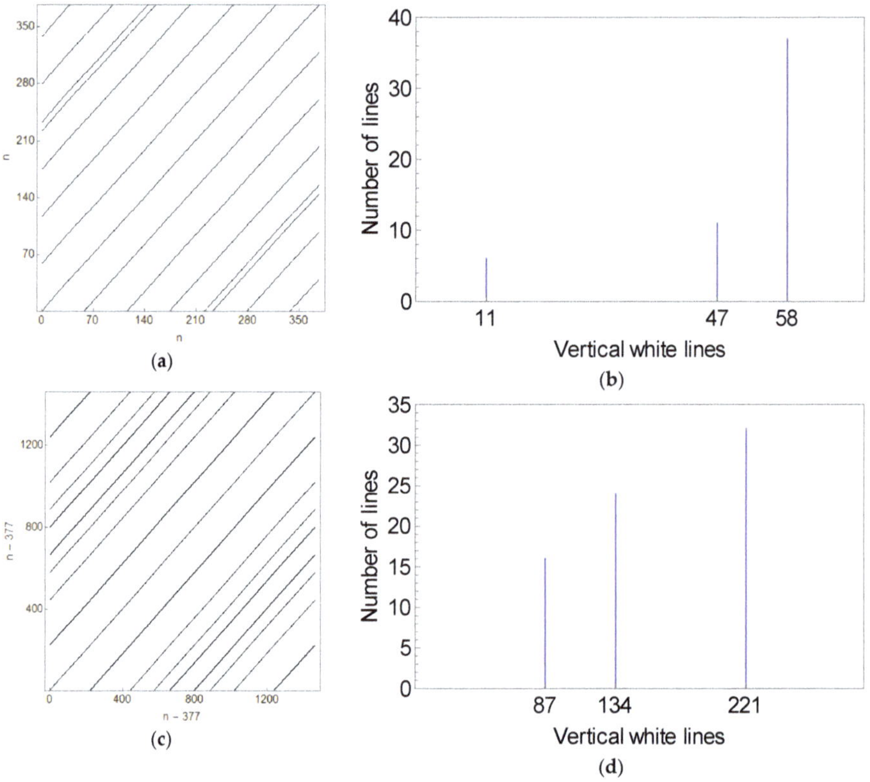

Figure 14. Two first block-like structures observed in the RP of Figure 8a. (**a**,**c**) Recurrence plots, $\varepsilon = 0.01$ and $\varepsilon = 0.003$, respectively. (**b**,**d**) Corresponding histograms of vertical white lines in the RP. A vertical white line denotes a calculated recurrence time. The number of lines is the number of times that such a recurrence time was observed.

The particle in the near singular billiard behaves quasi-periodically, filling a straight line in the phase space and possessing three return times, until it collides with the scatterer. When this happens, the angular momentum of the particle is changed, another line in the space starts to be filled, and the other three return times are measured on the new torus. The whole phase space is ergodically filled by one single trajectory which has a constant angular momentum between the two collisions with the scatterer. This explains the structures observed in the RP, where the bigger the structure, the longer the time the particle has spent in a given torus.

Sensitivity dependence on initial conditions was also verified for orbits of the near singular billiard. To exemplify this behavior, we set an initial condition in a very close neighborhood of the one used for Figure 13, in such a way that a distance $\delta_0 = 10^{-10}$ initially separates each orbit in the phase space, i.e., $(L_0, S_0) = (0.41 + \delta_0/\sqrt{2},\ 0.42 + \delta_0/\sqrt{2})$. Figure 15a shows the corresponding RP and Figure 15b shows a histogram of vertical white lines.

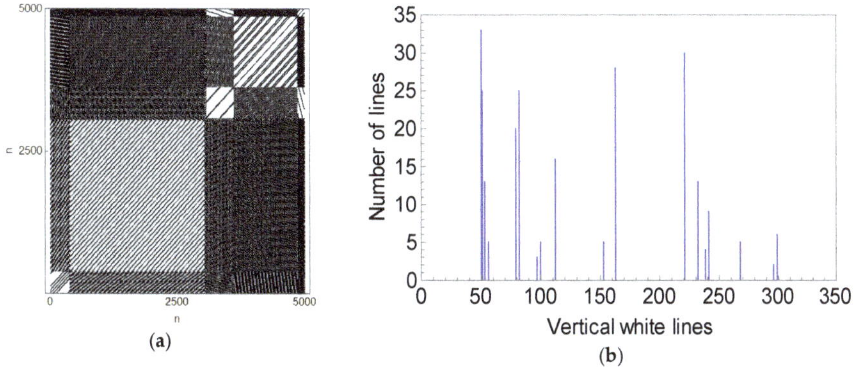

Figure 15. Orbit in the near singular billiard neighbor to $(L_0, S_0) = (0.41,\ 0.42)$, separated by a distance $\delta_0 = 10^{-10}$, and number of iterations $n = 5 \times 10^3$. (**a**) Recurrence plot, $\varepsilon = 0.006$, showing a junction of two quasi-periodic orbits. (**b**) Corresponding histogram of vertical white lines in the RP. A vertical white line stands for a calculated recurrence time and the number of lines is the number of times that such recurrence time was observed.

The RPs of Figures 13a and 15a show the same qualitative properties. Both indicate a chaotic orbit composed of quasi-periodic segments. The difference is in the size of the block-like structures in the diagonals. This is because a small change in the initial condition makes the particle interact with the scatterer at a different time and go to a different torus. The bigger the size of the block-like structure in the RP, the longer the particle has spent in the corresponding torus. Both initial conditions densely fill the phase space with straight lines, but the tori are visited in a different order. Evidently the recurrence times calculated for these two cases also differ, as seen in Figures 13b and 15b.

4. Discussion

We have shown that a typical orbit in the near singular limit of the annular billiard, described by a two-dimensional mapping, densely fills the phase space with straight lines. In order to characterize the dynamics as regular or chaotic, two different methods were applied: The traditional estimation of the largest Lyapunov exponent with the FTLE method and an analysis of recurrence properties through the recurrence plots.

The results obtained with the Lyapunov exponent analysis indicate that the dynamical behavior of the system changes when the near singular limit is reached, since a sharp drop in the value of λ occurs. The RPs provide more information for a better characterization of the dynamics at this limit, where they show that the orbit is chaotic but composed of quasi-periodic segments. This agrees with [36], which postulates on one-dimensional chaotic systems, and here, we have verified for a two-dimensional chaotic system: it is not possible to remove the chaotic behavior of chaotic nonlinear systems by reducing the magnitude of the nonlinear part.

Quantitative results on this, and possibly in other models, can be obtained through recurrence quantification analysis. It is also of interest to verify if a breathing version of this near singular billiard will hold true for the Fermi acceleration mechanism. Other billiards with scatterers can also be analyzed in the near singular limit.

Author Contributions: Conceptualization, R.E.d.C.; Data curation, R.S.B.; Formal analysis, R.S.B., B.C., R.E.d.C. and B.F.; Funding acquisition, R.E.d.C.; Investigation, R.S.B.; Methodology, R.E.d.C.; Project administration, R.E.d.C.; Resources, R.E.d.C.; Software, R.S.B.; Supervision, R.E.d.C.; Validation, B.C. and B.F.; Visualization, R.S.B.; Writing—original draft, R.S.B.; Writing—review and editing, R.E.d.C., B.C. and B.F.

Funding: We acknowledge support from the Brazilian scientific agencies CAPES—Coordination for the Improvement of Higher Education Personnel, CNPQ—National Council for Scientific and Technological Development (grant 306034/2015-8) and FAPESP—São Paulo Research Foundation.

Conflicts of Interest: The authors declare no conflict of interest.

References

1. Birkhoff, G.D. *Dynamical Systems*, 2nd ed.; American Mathematical Society: New York, NY, USA, 1927; pp. 154–196.
2. Sinai, Y.G. Dynamical systems with elastic reflections. *Rus. Math. Surv.* **1970**, *25*, 137–188. [CrossRef]
3. Bunimovich, L.A. On the ergodic properties of nowhere dispersing billiards. *Commun. Math. Phys.* **1979**, *75*, 295–312. [CrossRef]
4. Friedman, N.; Kaplan, A.; Carasso, D.; Davidson, N. Observation of Chaotic and Regular Dynamics in Atom-Optics Billiards. *Phys. Rev. Lett.* **2001**, *86*, 1518. [CrossRef] [PubMed]
5. Artigue, A. Billiards and Toy Gravitons. *J. Stat. Phys.* **2019**, *3*, 1–20. [CrossRef]
6. Palmeiro, M.S.; Livorati, A.L.P.; Caldas, I.L.; Leonel, E.D. Ensemble separation and stickiness influence in a driven stadium-like billiard: A Lyapunov exponents analysis. *Commun. Nonlinear Sci. Numer. Simul.* **2018**, *65*, 248–259. [CrossRef]
7. Zou, Y.; Pazó, D.; Romano, M.C.; Thiel, M.; Kurths, J. Distinguishing quasiperiodic dynamics from chaos in short-time series. *Phys. Rev. E* **2007**, *76*, 016210. [CrossRef]
8. Karney, C.F.F. Long-time correlations in the stochastic regime. *Physica D* **1983**, *8*, 360–380. [CrossRef]
9. Meiss, J.D.; Ott, E. Markov-Tree Model of Intrinsic Transport in Hamiltonian Systems. *Phys. Rev. Lett.* **1985**, *55*, 2741. [CrossRef]
10. Leoncini, X.; Zaslavsky, G.M. Jets, stickiness, and anomalous transport. *Phys. Rev. E* **2002**, *65*, 046216. [CrossRef]
11. Benettin, G.; Galgani, L.; Giorgilli, A.; Galdani, L.; Strelcyn, J.M. Lyapunov characteristic exponents for smooth dynamical systems and for Hamiltonian systems; A method for computing all of them. Part 2: Numerical application. *Meccanica* **1980**, *15*, 21–30. [CrossRef]
12. Zaslasvsky, G.M. *Physics of Chaos in Hamiltonian Systems*; Imperial College Press: London, UK, 2007.
13. Szezech, J.D., Jr.; Lopes, S.R.; Viana, R.L. Finite-time Lyapunov spectrum for chaotic orbits of non-integrable Hamiltonian systems. *Phys. Lett. A* **2005**, *335*, 394–401. [CrossRef]
14. Manchein, C.; Beims, M.W.; Rost, J.M. Characterizing weak chaos in non-integrable Hamiltonian systems: The fundamental role of stickiness and initial conditions. *Physica A* **2014**, *400*, 186–193. [CrossRef]
15. Szezech, J.D.; Lopes, S.R.; Viana, R.L. Finite-time Lyapunov spectrum for chaotic orbits of non-integrable Hamiltonian systems. *Phys. Lett. A* **2015**, *335*, 394–401. [CrossRef]
16. Slater, N. Gaps and steps for the sequence $n\theta$ mod 1. *Proc. Camb. Philos. Soc.* **1967**, *63*, 1115–1123. [CrossRef]
17. Altmann, E.G.; Cristadoro, G.; Pazó, D. Nontwist non-Hamiltonian Systems. *Phys. Rev. E* **2006**, *73*, 056201. [CrossRef] [PubMed]
18. Recurrence Plots and Cross Recurrence Plots. Available online: http://www.recurrence-plot.tk/glance.php (accessed on 25 February 2019).
19. Marwan, N. A historical review of recurrence plots. *Eur. Phys. J. Spec. Top.* **2008**, *164*, 3–12. [CrossRef]
20. Hoshi, R.A.; Pastre, C.M.; Vanderlei, L.C.M.; Godoy, M.F. Assessment of Heart Rate Complexity Recovery from Maximal Exercise Using Recurrence Quantification Analysis. In Proceedings of the 6th International Symposium on Recurrence Plots, Grenoble, France, 17–19 June 2015.
21. Arce, H.; Fuentes, A.; González, G.H. Recurrence Analysis of Cardiac Restitution in Human Ventricle. In Proceedings of the 6th International Symposium on Recurrence Plots, Grenoble, France, 17–19 June 2015.
22. Rizzi, M.; Frigerio, F.; Iori, V. The Early Phase of Epileptogenesis Induced by Status Epilepticus Are Characterized by Persistent Dynamical Regime of Intermittency Type. In Proceedings of the 6th International Symposium on Recurrence Plots, Grenoble, France, 17–19 June 2015.
23. Flach, M.; Lange, H.; Foken, T.; Hauhs, M. Recurrence Analysis of Eddy Covariance Fluxes. In Proceedings of the 6th International Symposium on Recurrence Plots, Grenoble, France, 17–19 June 2015.
24. Kabiraj, L.; Saurabh, A.; Nawroth, H.; Paschereit, C.O.; Sujith, R.I.; Karimi, N. Recurrence Plots for the Analysis of Combustion Dynamics. In Proceedings of the 6th International Symposium on Recurrence Plots, Grenoble, France, 17–19 June 2015.

25. Viana, R.L.; Toufen, D.L.; Guimarães-Filho, Z.O.; Caldas, I.L.; Gentle, K.W.; Nascimento, I.C. Recurrence Analysis of Turbulent Fluctuations in Magnetically Confined Plasmas. In Proceedings of the 6th International Symposium on Recurrence Plots, Grenoble, France, 17–19 June 2015.
26. Zou, Y.; Thiel, M.; Romano, M.C.; Kurths, J. Characterization of stickiness by means of recurrence. *Chaos* **2007**, *17*, 043101. [CrossRef] [PubMed]
27. Seba, P. Wave Chaos in Singular Quantum Billiard. *Phys. Rev. Lett.* **1990**, *64*, 1855–1858. [CrossRef] [PubMed]
28. Bohigas, O.; Boosé, D.; Carvalho, R.E.; Marvulle, V. Quantum tunneling and chaotic dynamics. *Nucl. Phys. A* **1993**, *560*, 197–210. [CrossRef]
29. Cheon, T.; Shigehara, T. Geometric Phase in Quantum Billiards with a Pointlike Scatterer. *Phys. Rev. Lett.* **1996**, *76*, 1770–1773. [CrossRef] [PubMed]
30. Carvalho, R.E.; Souza, F.C.; Leonel, E.D. Fermi acceleration on the annular billiard. *Phys. Rev. E* **2006**, *73*, 066229. [CrossRef] [PubMed]
31. Carvalho, R.E.; Souza, F.C.; Leonel, E.D. Fermi acceleration on the annular billiard: A simplified version. *J. Phys. A* **2006**, *39*, 3561–3573. [CrossRef]
32. Altmann, E.G.; Friedrich, T.; Motter, A.E.; Kantz, H.; Richter, A. Prevalence of marginally unstable periodic orbits in chaotic billiards. *Phys. Rev. E* **2008**, *77*, 016205. [CrossRef] [PubMed]
33. Altmann, E.G.; Leitão, J.C.; Lopes, J.V. Effect of noise in open chaotic billiards. *Chaos* **2012**, *22*, 026114. [CrossRef]
34. Abud, C.V.; Carvalho, R.E. Multifractality, stickiness, and recurrence-time statistics. *Phys. Rev. E* **2013**, *88*, 042922. [CrossRef] [PubMed]
35. Dettmann, C.P.; Fain, V. Linear and nonlinear stability of periodic orbits in annular billiards. *Chaos* **2017**, *27*, 043106. [CrossRef] [PubMed]
36. Katz, L. Weak Chaos. *Chaos Solitons Fractals* **1995**, *7*, 1057–1063. [CrossRef]

Mathematical and Computational Applications

MDPI

Article

Functional Ca^{2+} Channels between Channel Clusters are Necessary for the Propagation of IP_3R-Mediated Ca^{2+} Waves

Estefanía Piegari and Silvina Ponce Dawson *

Departamento de Física FCEN-UBA and IFIBA (CONICET), Ciudad Universitaria, Pabellón I, Buenos Aires 1428, Argentina; estefipiegari@gmail.com

* Correspondence: silvina@df.uba.ar

Received: 10 May 2019; Accepted: 10 June 2019; Published: 11 June 2019

Abstract: The specificity and universality of intracellular Ca^{2+} signals rely on the variety of spatio-temporal patterns that the Ca^{2+} concentration can display. Ca^{2+} release into the cytosol through inositol 1,4,5-trisphosphate receptors (IP_3Rs) is key for this variety. The opening probability of IP_3Rs depends on the cytosolic Ca^{2+} concentration. All of the dynamics are then well described by an excitable system in which the signal propagation depends on the ability of the Ca^{2+} released through one IP_3R to induce the opening of other IP_3Rs. In most cell types, IP_3Rs are organized in clusters, i.e., the cytosol is a "patchy" excitable system in which the signals can remain localized (i.e., involving the release through one or more IP_3Rs in a cluster), or become global depending on the efficiency of the Ca^{2+}-mediated coupling between clusters. The spatial range over which the signals propagate determines the responses that the cell eventually produces. This points to the importance of understanding the mechanisms that make the propagation possible. Our previous qualitative comparison between experiments and numerical simulations seemed to indicate that Ca^{2+} release not only occurs within the close vicinity of the clearly identifiable release sites (IP_3R clusters) but that there are also functional IP_3Rs in between them. In this paper, we present a quantitative comparison between experiments and models that corroborate this preliminary conclusion. This result has implications on how the Ca^{2+}-mediated coupling between clusters works and how it can eventually be disrupted by the different Ca^{2+} trapping mechanisms.

Keywords: calcium signals; IP_3Rs dsitribution; puffs; waves

1. Introduction

Calcium (Ca^{2+}) signals are ubiquitous across cell types [1,2]. In many cases, they involve Ca^{2+} release from the endoplasmic reticulum (ER) into the cytosol through Inositol 1,4,5-trisphosphate receptors (IP_3Rs), which are Ca^{2+} channels [3]. The opening probability of IP_3Rs increases with the cytosolic Ca^{2+} concentration [4], provided that this concentration is not too high. Thus, the very same Ca^{2+} that is released through an open IP_3R can induce the opening of neighboring IP_3Rs. The combination of this *Calcium Induced Calcium Release* (CICR) [5] with the diffusion of Ca^{2+} between its channels gives rise to propagating signals that can even embrace the whole cell [6–8]. IP_3Rs become inhibited in the presence of high cytosolic Ca^{2+} concentrations. From a physical/mathematical point, the dynamics that underlie these propagating Ca^{2+} signals are well described by an excitable system. Now, in most cell types, IP_3Rs are organized in clusters. Waves can then fail to propagate if the amount of Ca^{2+} that reaches one cluster is not high enough to "cross" the excitability threshold [9]. We have recently studied the Ca^{2+}-mediated coupling between neighboring IP_3R-clusters by means of experiments in which we used two single-wavelength Ca^{2+}-dyes [10]. Single-wavelength Ca^{2+}-dyes are Ca^{2+} indicators that increase their fluorescence enormously upon Ca^{2+} binding, without changing their emission wavelength [11]. The presence of the dyes affects the elicited signals because they "trap" Ca^{2+} and, in this way, can disrupt CICR. How the presence of different Ca^{2+} trapping mechanisms (particularly, Ca^{2+}-binding *buffers*) affects Ca^{2+} signals had been studied with experiments in which varying quantities of exogenous Ca^{2+} buffers had been introduced in the cells [12,13]. The experiments of [10] allowed us to make visible the invisible: by observing the signals simultaneously with a slow (Rhod-2) and a fast (Fluo-4) dye, we could infer directly the different ways in which slow and fast buffers affect the signals. Furthermore, performing experiments for various concentrations of the dyes and comparing qualitatively the spatio-temporal distributions of the Ca^{2+}-bound to both dyes with those derived from numerical simulations, we inferred that there should be functional IP_3Rs between the IP_3R-clusters to explain the observations. This implied that Ca^{2+} release not only occurred within the close vicinity of the centers of the clearly identifiable release sites (the clusters) but also in between.

In this paper, we present a more quantitative analysis of the experimental observations. To this end, we compute, from the experiments, the change with the slow dye concentration, Rhod-2, of the probability, $P(\ell \mid n_c)$ that, given that there are n_c simultaneously open channels in a *primary* cluster, the event does not induce the opening of IP_3Rs in other (*secondary*) clusters (i.e., the event remains localized). We then use a simple model to compute numerically the probability that one IP_3R located at a distance, d, from a primary IP_3R-cluster becomes open (after a certain time) given that there are n_c simultaneously open channels at the primary cluster. Changing the parameters of the simulation we conclude that the changes observed experimentally can only be explained if $d \sim 0.6$ µm (i.e., it is smaller than the typical inter-cluster distance, $d \sim 1.4$ µm) and that the basal Ca^{2+} concentration, $[Ca]_b$, is reduced when the slow dye concentration, $[R]_T$, is increased.

We present in what follows the experimental results that we use to estimate the changes in $P(\ell \mid n_c)$ as $[R]_T$ is varied. We introduce in Section 2.2 the probabilistic model that we will then use to analyze the experimental data. In Section 2.3, we show the results of the numerical simulations with which we estimate some of the probabilities that enter the probabilistic model. In Section 2.4, we combine the experimental and numerical results and determine for what parameters of the simulations they are compatible within the framework probabilistic model. A discussion is included at the end.

2. Results

2.1. Experimental Results

The experiments analyzed here were presented and described in detail in [10]. A description of how they were performed is included in Materials and Methods. In brief, we elicited IP_3-mediated Ca^{2+} release events in *Xenopus laevis* oocytes that were previously injected with caged IP_3, the Ca^{2+} buffer, EGTA, and a fast, Fluo-4, and a slow, Rhod-2, Ca^{2+} dye. The signals were elicited by uncaging the caged IP_3 with an UV flash. In the current paper, we analyze the changes observed in the distribution of localized Ca^{2+} release events (*puffs*) elicited in this way as the concentration of the slow dye, $[R]_T = [Rhod{-}2]$, is varied but all other experimental parameters are kept fixed ($[EGTA] = 90$ µM; $[Fluo{-}4] = [F]_T = 36$ µM, duration of the UV flash to uncage the $IP_3 = (100 - 200)$ ms; see Table 1). In particular, we are interested in studying the changes in the "size" of the elicited release events that remain localized (Ca^{2+} puffs), where by size we mean the number of IP_3Rs that are simultaneously open at the release site (the cluster) during the release event. In order to compare the localized event size distributions obtained in experiments performed for different values of $[R]_T$, we introduced in [10] a quantity, A_{lib-F}, that is an increasing function of the Ca^{2+} current that underlies the observed release event regardless of the value of $[R]_T$. We describe in Section 4.3 how we compute A_{lib-F} from the fluorescence emitted by the Ca^{2+}-bound Fluo-4 molecules. We show in Figure 1 the cumulative distribution functions (CDFs) of A_{lib-F} derived from the experiments performed for the conditions of Set **III** (dashed line), Set **II** (dotted line) and Set **I** (solid line). These CDFs were computed including only localized events, i.e., Ca^{2+} puffs.

Table 1. Combinations of the dyes and EGTA concentrations used in the different experiments.

Experiment	[Fluo-4] (μM)	[Rhod-2] (μM)	[EGTA] (μM)
Set **I**	36	90	90
Set **II**	36	36	90
Set **III**	36	0	90

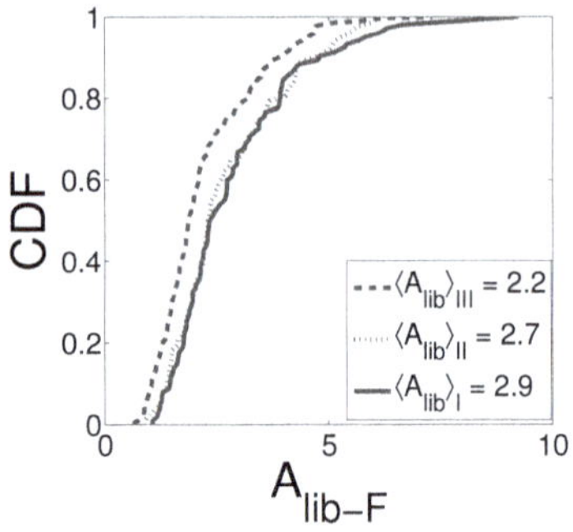

Figure 1. CDF of event sizes (as measured by A_{lib-F}) for experiments performed with $[F]_T = 36$ μM and $[EGTA] = 90$ μM and different values of $[R]_T$ (the solid line corresponds to Set **I**, the dotted line to Set **II** and the dashed line to Set **III** (see Table 1).

The Kolmogorov–Smirnov test rejects the null hypothesis that A_{lib-F} from sets **III** and **II** and sets **III** and **I** come from the same continuous distribution with a 99% significance level ($p_{value} = 2.1 \times 10^{-4}$ and $p_{value} = 4.6 \times 10^{-5}$, respectively), but cannot reject that the data points from Set **II** and Set **I** come from the same distribution ($p_{value} = 0.96$). In any case, there is a tail in the CDF of A_{lib-F} for Set **I** that is unobservable in that of Set **II** which is consistent with having more events with relatively larger underlying Ca^{2+} currents in the former than in the latter ($\langle A_{lib-F} \rangle = 2.2$ for Set **III** $\langle A_{lib-F} \rangle = 2.7$ for Set **II** and $\langle A_{lib-F} \rangle = 2.9$ for Set **I**). These comparisons indicate that puffs with relatively larger underlying Ca^{2+} currents can be elicited as the concentration of the slow dye, Rhod-2, is increased. Although the transformation from puff amplitude to A_{lib-F} involves certain uncertainties, the changes observed in the fluorescence rise time as $[R]_T$ is varied [10] support this conclusion.

2.2. Probabilistic Model to Analyze the Differences Observed in the Experimental Event Size Distributions for Different Values of $[R]_T$

As analyzed in [14], being able to observe localized Ca^{2+} release events (puffs) with larger Ca^{2+} currents as a slow buffer concentration (in this case, Rhod-2) is increased can be due to a more efficient uncoupling between IP_3R clusters due to the presence of the slow buffer. Namely, we have the hypothesis that the differences in the CDFs of localized release event sizes illustrated in Figure 1 occur because, as $[R]_T$ decreases, Ca^{2+} release events with too many simultaneously open IP_3Rs at the primary site can no longer remain localized, induce the opening of IP_3Rs in neighboring (secondary) clusters and, thus, are not included to compute the CDF. We hereby introduce a way to analyze the experimental data to quantify what fraction of events that are localized for a given $[R]_T$ turn into waves as $[R]_T$ is decreased.

We define $P(n_c)$ as the probability that there are n_c simultaneously open IP_3Rs in a cluster for a given set of experimental conditions. Here, we will assume that all conditions remain the same except for the total slow dye concentration, $[R]_T$. Thus, we will analyze the change of $P(n_c)$ with $[R]_T$. Given that there are n_c simultaneously open IP_3Rs we want to distinguish whether this situation induces the opening of at least one IP_3R in a neighboring (secondary) cluster (i.e., it initiates a wave) or it does not (i.e., the Ca^{2+} release event due to the n_c simultaneously open IP_3Rs remains localized). We then write:

$$P(n_c) = P(n_c \,\&\, \ell \mid [\mathrm{R}]_T) + P(n_c \,\&\, w \mid [\mathrm{R}]_T). \tag{1}$$

In Equation (1), $P(n_c \,\&\, \ell \mid [\mathrm{R}]_T)$ is the joint probability that n_c channels are simultaneously open in a cluster and the event stays localized for a given value of $[\mathrm{R}]_T$. $P(n_c \,\&\, w \mid [\mathrm{R}]_T)$ is the joint probability that n_c channels are simultaneously open in a cluster and the event induces the opening of at least one IP_3R in another cluster for a given value of $[\mathrm{R}]_T$. The symbol $\mid$ means that these are two conditional probabilities for a given value of the slow dye concentration, $[R]_T$. All the probabilities we work with here are defined over the set of events, i.e., for $n_c \geq 1$. The aim of this calculation is to assess how the two joint probabilities of Equation (1) change with $[R]_T$. Under the assumption that Rhod-2 is a slow buffer and, as such, does not affect CICR within the cluster [14], we consider that $P(n_c)$ does not depend on $[R]_T$. What may change when varying Rhod-2 is whether the event with n_c open channels in a cluster remains localized (stays as a puff) or elicits the opening of channels in a neighboring cluster (becomes a wave). We rewrite the two joint probabilities of interest as:

$$\begin{aligned} P(n_c \,\&\, \ell \mid [\mathrm{R}]_T) &= P(n_c \mid \ell, [\mathrm{R}]_T) P(\ell \mid [\mathrm{R}]_T) \\ &= P(n_c \mid \ell, [\mathrm{R}]_T)\left(1 - P(w \mid [\mathrm{R}]_T)\right), \end{aligned} \tag{2}$$

$$P(n_c \,\&\, w \mid [\mathrm{R}]_T) = P(w \mid n_c, [\mathrm{R}]_T) P(n_c). \tag{3}$$

In these equations, $P(n_c \mid \ell, [\mathrm{R}]_T)$ is the probability that a Ca^{2+} release event that remains localized for a given $[\mathrm{R}]_T$ corresponds to a situation with n_c simultaneously open channels at the release site; $P(w \mid n_c, [\mathrm{R}]_T)$ is the probability that, for n_c open channels in a cluster and a given $[\mathrm{R}]_T$, the resulting event induces the release of Ca^{2+} from a neighboring cluster (i.e., generates a wave). $P(\ell \mid n_c, [\mathrm{R}]_T)$ and $P(w \mid n_c, [\mathrm{R}]_T)$, on the other hand, are the probabilities that a Ca^{2+} release event obtained for given $[\mathrm{R}]_T$ remains localized or initiates a wave, respectively. They satisfy: $P(\ell \mid n_c, [\mathrm{R}]_T) + P(w \mid n_c, [\mathrm{R}]_T) = 1$. We rewrite the latter as:

$$P(w \mid [\mathrm{R}]_T) = \sum_{n_c \geq 1} P(w \mid n_c, [\mathrm{R}]_T) P(n_c). \tag{4}$$

Combining Equations (1)–(3), we arrive at:

$$\begin{aligned} P(n_c) &= P(n_c \mid \ell, [\mathrm{R}]_T)\left(1 - P(w \mid [\mathrm{R}]_T)\right) \\ &+ P(w \mid n_c, [\mathrm{R}]_T) P(n_c). \end{aligned} \tag{5}$$

Assuming that the Ca^{2+} current through an open IP_3R is approximately the same for all IP_3Rs, we conclude that n_c is proportional to the Ca^{2+} current that underlies a Ca^{2+} release event. The quantity A_{lib-F} that we derive from the experimental data, on the other hand, is an increasing function of the underlying Ca^{2+} current. We must point out that, if n_c is large enough, puff amplitudes increase sublinearly with n_c [15]. Assuming that n_c and A_{lib-F} are approximately linearly related, we can then use the experimental CDF of A_{lib-F}, which we compute for the localized Ca^{2+} release events, to estimate the CDF, F that can be computed from $P(n_c \mid \ell, [\mathrm{R}]_T)$:

$$F(n, \mid \ell, [R]_T) = \sum_{n_c=1}^{n} P(n_c \mid \ell, [R]_T). \tag{6}$$

The aim is to compare the distribution functions, $F(n, \mid \ell, [R]_T)$, for different values of $[R]_T$ using the corresponding experimental CDFs of A_{lib-F}. In particular, we will compare the CDFs that are sufficiently different according to the K–S test: the ones with $[R]_T = 0$ and with $[R]_T = 90$ μM. In what follows, we will drop the concentration units (μM) from the expressions of the probabilities to simplify the notation. Defining $\Delta P_w(n_c) \equiv P(w\,|\,n_c, 0) - P(w\,|\,n_c, 90)$ and $\Delta P_w \equiv \sum_{n_c \geq 1} \Delta P_w(n_c) P(n_c)$ and using Equations (4) and (5), we obtain:

$$\Delta P_w(n_c)P(n_c) = (P(n_c\,|\,\ell, 90) - P(n_c\,|\,\ell, 0))\,(1 - P(w\,|\,90)) + P(n_c\,|\,\ell, 0)\Delta P_w. \quad (7)$$

As illustrated in Figure 1, the experiments show that the difference between the two CDFs is more noticeable in the region of the largest size events, i.e., for the largest values of n_c. We then compute:

$$\sum_{n_c \geq n_M} \Delta P_w(n_c)P(n_c) = (F(n_M\,|\,\ell, 0) - F(n_M\,|\,\ell, 90))\,(1 - P(w\,|\,90)) + \Delta P_w(1 - F(n_M\,|\,\ell, 0)), \quad (8)$$

where n_M is the event size beyond which the CDFs start to differ more noticeably. As described later, the CDFs in the r.h.s. of Equation (8) can be estimated from the experimental CDFs. On the other hand, we estimate $P(w\,|\,n_c, [\mathrm{R}]_T)$ using the numerical simulations that we describe in the following section. Varying the parameters of the simulation, we determine the values for which we obtain estimates of the l.h.s. of this equation that are consistent with those of the r.h.s.

2.3. Numerical Simulations to Estimate the Probability That a Release Event from One (Primary) Cluster Induces the Release of Ca^{2+} from Another (Secondary) Cluster

We compute the probability, $P_0(t, d, n_c, n_s, [R]_T)$, that n_s IP_3R located at a distance, d, from a (primary) cluster with n_c IP_3Rs that are simultaneously open at $t = 0$, becomes open by a time, t. We want to compare how P_0 varies as $[R]_T$ is changed. We thus write explicitly its dependence on this variable. To compute $P_0(t, d, n_c, n_s, [R]_T)$, we proceed as explained in Materials and Methods (see also [10]) and the parameter values used are listed in Table 2. We show in Figure 2 the results obtained with $n_s = 1$. We show in Figure 2a–c the results obtained using the basal Ca^{2+} concentration, $[\mathrm{Ca}]_b$= 0.1 μM, for the conditions of Sets **I**, **II** and **III**. We show the results obtained at $d = 0.6$ μm in Figure 2a and at $d = 1.4$ μm (a typical inter-cluster distance) in Figure 2b,c. The number of simultaneously open channels is $n_c = 10$ in Figure 2a,b and $n_c = 50$ in Figure 2c. The change of P_0 with varying $[R]_T$ is unobservable for $n_c = 10$ at $d = 1.4$ μm (the difference is ≤ 0.004 for the times displayed in the figure) while it can be ~ 0.085 at $d = 0.6$ μm. Furthermore, it is $\Delta P_0(t) = P_0(t, d = 0.6\ \mu\mathrm{m}, n_c = 10, [R]_T = 90\ \mu\mathrm{M}) - P_0(t, d = 0.6\ \mu\mathrm{m}, n_c = 10, [R]_T = 0\ \mu\mathrm{M}) \approx 0.065$ at $t = d/V$ with $V \sim 10\ \mu\mathrm{ms}^{-1}$, a typical wave velocity. The maximum difference $\max_t \Delta P_0(t)$ increases with n_c. This is shown in Figure 2c where $n_c = 50$, $d = 1.4$ μm and $\max_t \Delta P_0(t) \sim 0.026$. In Figure 2d, we show what happens when $[\mathrm{Ca}]_b$ decreases. In this case, we compare P_0 at a distance $d = 1.4$ μm from the source obtained for simulations performed with the concentrations of Set **III** (dashed line) and Set **I** (solid line) but with a different value of $[\mathrm{Ca}]_b$ in each one (100 nM and 50 nM, respectively). A similar behaviour is obtained with $n_s = 5$ (data not shown).

Table 2. Value of the parameters varied to compute P_0.

Parameter	Abbreviature	Values
Number of IP_3Rs in the source	n_c	1, 10, 50
Distance to the Ca^{2+} source	d	(0.4-1.5) μm
Number of sensing IP_3Rs	n_s	1, 5
Rhod-2 concentration	$[R]_T$	0, 90 μM
Velocity of propagation	V	10, 20 μm/s

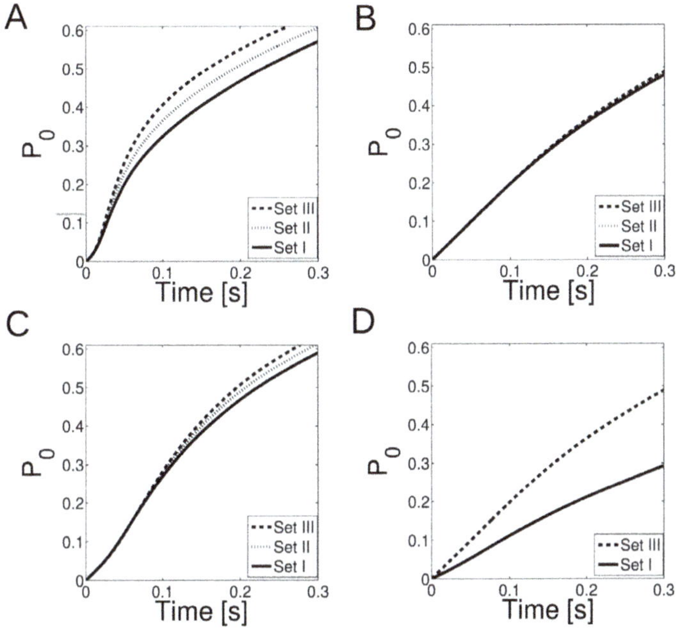

Figure 2. P_0 dependence with distance d from the source and $[\mathrm{Ca}]_b$. (**a**) performed at $d = 0.6$ µm with $n_c = 10$; (**b**) at $d = 1.4$ µm with $n_c = 10$; (**c**) at $d = 1.4$ µm and $n_c = 50$ and (**d**) $d = 1.4$ µm with $n_c = 10$ while varying $[\mathrm{Ca}]_b$ from 100 nM (Set **III**) to 50 nM (Set **I**). The dashed line corresponds to Set **III**, the dotted line corresponds to Set **II**, and the solid line corresponds to Set **I**.

We interpret the $[R]_T$-dependent changes of P_0 that are illustrated in Figure 2 as a sign of the change in the level of inter-cluster coupling (or, equivalently, disruption) that can be reached as the slow dye concentration is varied.

We further studied how sensitive is P_0 to changes in the distance to the source d. The results of Figure 3 are obtained using the basal Ca^{2+} concentration, $[\mathrm{Ca}]_b$= 0.1 µM. To illustrate the disruption when $[R]_T$ is increased, we show in Figure 3a $\Delta P_0 = P_0([R]_T = 0) - P_0([R]_T = 90\ \mu\mathrm{M})$ computed with $n_c = 10$ and $n_s = 1$ as a function of the distance d for each time $t = d/V$ (with $V = 10$ µm/s) and it can be observed that the probability of opening one IP_3R in-between clusters decreases (from 0.4 µm to 1.5 µm). When observing the probability of opening one IP_3R as a function of d without adding the slow buffer ($R_T = 0$, Set **III**) (solid line in Figure 3b), as d increases, this probability approximates to the basal probability (dotted line, P_0 computed as in Equation (14) but with no calcium dyes), almost no coupling can occur at the typical inter-cluster distance ($d = 1.4$ µm). Thus, to explain the inter-cluster coupling, it is necessary to add a non-cluster IP_3R in-between them. The optimal value of the parameter d should be on the order of 0.4–0.8 µm (approximately the half distance between clusters). Not even adding $n_s = 5$ sensing channels at the second cluster, the probability differs from the basal (dashed and dotted lines in Figure 3c, respectively). We choose $d = 0.6$ µm to add an isolated IP_3R in-between clusters (solid line in Figure 3c) and now the signal can propagate.

We now study whether the variations of Figure 2 can explain the changes in the distributions of Ca^{2+} release during localized events observed in the experiments that are apparent in Figure 1.

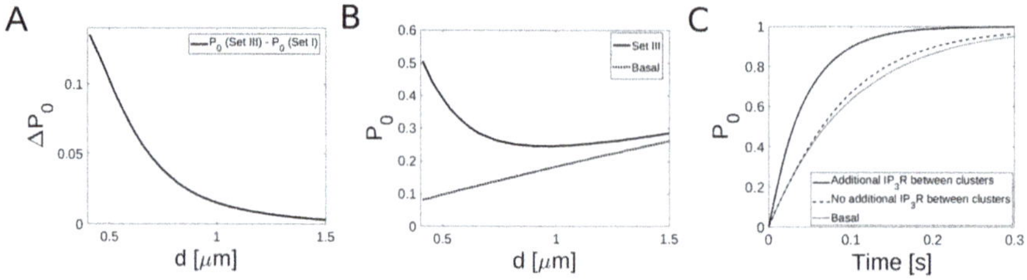

Figure 3. Existence of IP$_3$Rs in-between clusters is necessary to explain the observations. (**a**) $\Delta P_0 = P_0([R]_T = 0) - P_0([R]_T = 90\ \mu\text{M})$ with $n_c = 10$ and $n_s = 1$ as a function of the distance d for each time $t = d/V$ (with $V = 10\ \mu\text{m/s}$); (**b**) P_0 computed as in (**a**) for the Set **III** (solid line) and basal (dotted line) conditions; (**c**) P_0 computed in the condition of Set **III** with $n_c = 10$ and $n_s = 5$ at a distance $d = 1.4\ \mu\text{m}$ from the Ca^{2+} point source as a function of time when an IP$_3$R is added in-between cluster (solid line), with no additional IP$_3$R (dashed line) and at the basal condition (dotted lined).

2.4. Combining the Estimates Derived from the Experiments and from the Numerical Simulations to Interpret the Changes Observed Experimentally

We first estimate the r.h.s. of Equation (8) assuming that $F(n, \mid \ell, [R]_T)$ is given by the experimental CDF of A_{lib-F} for the same $[R]_T$ and some unknown factor between n and A_{lib-F}. We recall here that the experimental CDF corresponds only to localized events (i.e., events at a primary cluster that do not induce the opening of IP$_3$Rs at another secondary IP$_3$R cluster). In order to estimate this unknown factor, we associate, n_M, (the value after which the differences in the CDFs become more noticeable) to a value, A_{lib-F}, for which $F(n_M \mid \ell, 0)$ is sufficiently close to 1. The basic assumption here is that, for $[R]_T = 0$, almost all primary events with $n \leq n_M$ simultaneously open IP$_3$Rs initiate waves in which case they do not remain localized and are, therefore, not included in the computation of the CDF. We choose $A_{lib-F} = 4.8$ for which, according to the experimental data, it is $F(n_M \mid \ell, 0) \approx 0.98$ (see Figure 1). For this value, it is $F\,(n_M \mid \ell, 90)) \approx 0.90$. Thus, we estimate $F(n_M \mid \ell, 0) - F(n_M \mid \ell, 90) \approx 0.08$. We do not have a direct estimation of $P(w \mid [\mathrm{R}]_T)$. Assuming that $n_c = 10$ is the most probable value for the number of simultaneously open IP$_3$Rs in a cluster, we approximate $P(w \mid [\mathrm{R}]_T) = \sum_{n_c \geq 1} P(w \mid n_c, [\mathrm{R}]_T) P(n_c) \approx P(w \mid n_c = 10, [\mathrm{R}]_T) \approx P_0(t = d/V, d, n_c = 10, [R]_T)$ with P_0 the open probability computed numerically that we introduced in the previous section and V a typical Ca^{2+} wave velocity. For V, we try two values, $V = 20\ \mu\text{m/s}$ and $V = 10\ \mu\text{m/s}$. For d, we try the typical inter-cluster distance, $d = 1.4\ \mu\text{m}$ and the closer distance, $d = 0.6\ \mu\text{m}$ that was probed in the previous section. Using $d = 1.4\ \mu\text{m}$ and $V = 20\ \mu\text{m/s}$, the simulations give $P(w \mid 90) \approx 0.14$ and $\Delta P_w \approx 0$. The estimate of the r.h.s. of Equation (8) then results equal to 0.07. This value changes to 0.06 if we use $V = 10\ \mu\text{m/s}$. Using $d = 0.6\ \mu\text{m}$ and $V = 20\ \mu\text{m/s}$ the simulations give $P(w \mid 90) \approx 0.12$ and $\Delta P_w = 0.028$. The estimate of the r.h.s. of Equation (8) then results as 0.07. This value changes to 0.06 if we use $V = 10\ \mu\text{m/s}$.

We now use the simulations of the previous section to put an upper bound, $\Delta P_{w,max}$, on $\Delta P_w(n_c)$ in the l.h.s of Equation (8). With such an upper bound, we can write $\sum_{n_c \geq n_M} \Delta P_w(n_c) P(n_c) \leq \Delta P_{w,max} \sum_{n_c \geq n_M} P(n_c) = \Delta P_{w,max}(1 - F(n_M))$ where $F(n_M) = \sum_{n_c \leq n_M} P(n_c)$. Similarly to the way we have followed estimating $P(w \mid [R]_T)$, we compute $P(w \mid n_c, [R]_T) \approx P_0(t = d/V, d, n_c, [R]_T)$, with P_0 the open probability of the previous section illustrated in Figure 2. In particular, using the results of these simulations, we conclude that $\Delta P_w(n_c) \equiv P(w \mid n_c, 0) - P(w \mid n_c, 90\ \mu\text{M})$ is larger the larger the value of n_c. Thus, we obtain the upper bound, $\Delta P_{w,max}$, using similar simulations to those of Figure 2 but for $n_c = 50$ (a very large number of simultaneously open IP$_3$Rs). Namely, we estimate $\Delta P_{w,max} \approx P_0(t = d/V, d, n_c = 50, 0) - P_0(t = d/V, d, n_c = 50, 90)$. In order to put an upper bound on the l.h.s. of Equation (8), we need a bound for $F(n_M)$, the CDF of all the (primary) event sizes at $n = n_M$. As already explained, we assume that $F(n) = \sum_{n_c \leq n} P(n_c)$ does not depend on $[R]_T$. Given our interpretation of the results, we assume that the difference between the CDF of all the (primary) event sizes, $F(n)$, and the CDF of the primary event sizes that remain localized for a given value of $[R]_T$, $F(n \mid \ell, [R]_T)$, is due to the existence of primary events (of large enough size) that initiate waves for that value of $[R]_T$. Assuming that the fraction of primary events that initiate waves for $[R]_T = 90$ μM is negligible, we can approximate $F(n_M) \approx F(n_M \mid \ell, 90) \approx 0.90$. If we do not want to use this approximation, then we can use the bound $F(n_M) < 0.90$. In what follows, we mostly use $F(n_M) = 0.90$, but we repeat some computations changing it to 0.80 to see how much the estimates could change. Proceeding as just explained, for $d = 1.4$ μm, we obtain $\Delta P_w(n_c) \leq \Delta P_{w,max} \approx 0.003$ for $V = 20$ μm/s and $\Delta P_{w,max} \approx 0.026$ for $V = 10$ μm/s. Using $F(n_M) = 0.9$, we then obtain ~ 0.0003 and 0.0026 as upper bounds of the l.h.s. of Equation (8) for $V = 20$ μm/s and $V = 10$ μm/s, respectively. These two upper bounds are at least one order of magnitude smaller than the values obtained for the r.h.s. of Equation (8). If we use $F(n_M) = 0.8$ to compute the l.h.s. of this equation, the latter estimate doubles with respect to the previous one. Thus, the order of magnitude difference between the left and right estimates for $d = 1.4$ μm remains the same. Repeating the computations for $d = 0.6$ μm, we obtain $\Delta P_w(n_c) \leq \Delta P_{w,max} \approx 0.19$ for $V = 20$ μm/s and $\Delta P_{w,max} \approx 0.15$ for $V = 10$ μm/s. Using $F(n_M) = 0.9$, we then get ~ 0.019 and 0.015 as upper bounds of the l.h.s. of Equation (8) for $V = 20$ μm/s and $V = 10$ μm/s, respectively. In this case, the values of the left- and right-hand sides are of the same order of magnitude. These estimates come closer together if we use $F(n_M) = 0.8$ in the l.h.s. of the equation. In such a case, we obtain ~ 0.04 and ~ 0.03 for the l.h.s. estimate using $V = 20$ μm/s and $V = 10$ μm/s, respectively, two values that are pretty similar to the r.h.s. estimates, 0.07 and 0.06.

2.5. Changes in Basal Calcium Concentration, $[Ca]_b$

As illustrated in Figure 2d, decreasing basal [Ca] with increasing $[R]_T$ changes the open probability at the distance, $d = 1.4$ μm, in the direction that is needed to explain the observed changes in the event size distributions. We analyze here whether there is any evidence of a decreasing basal Ca^{2+} with increasing $[R]_T$ in the experimental data. We show in Figure 4 the cumulative density functions of the mean basal fluorescence emitted by the Ca^{2+}-bound Fluo-4 molecules, $\langle f_{0,F} \rangle$, in (a) and of the mean basal fluorescence emitted by the Ca^{2+}-bound Rhod-2 molecules, $\langle f_{0,R} \rangle$ in (b) for the experiments with $[R]_T = 36$ μM (dotted line) and with $[R]_T = 90$ μM (solid line). In Figure 4b, we rescaled $\langle f_{0_R} \rangle$ by $90/36 = 2.5$ in the case of Set **I** to make the distributions of experiments of Set **I** (which has $[R]_T = 90$ μM) and **II** (which has $[R]_T = 36$ μM) readily comparable. The values of $\langle f_{0,F} \rangle$ and $\langle f_{0,R} \rangle$were derived from the fluorescence observations as explained in Materials and Methods. We observe that the CDFs move to smaller values of their arguments with increasing $[R]_T$. As the mean basal fluorescence is an increasing function of $[Ca]_b$ (see Equation (12)), this observation supports the idea that, on average, $[Ca]_b$ decreases with increasing Rhod-2.

In order to estimate the variation in $[Ca]_b$ with increasing $[R]_T$, we compare $\overline{\langle f_{0,D} \rangle}$ for sets **I** and **II**. In particular, we obtain $\overline{\langle f_{0,F} \rangle}$ = 6.1 a.u., $\overline{\langle f_{0,R} \rangle}$ = 13.2 a.u. and $\overline{\langle f_{0,F} \rangle}$ = 7.2 a.u., $\overline{\langle f_{0,R} \rangle}$ = 6.4 a.u. for sets **I** and **II**, respectively. Inserting these values into Equation (13), using that $\langle N_R \rangle = 32$ for set **II** and $\langle N_R \rangle = 80$ for set **I**, and assuming that $[Ca]_b$ = 100 nM for set **II**, we obtain $[Ca]_b$ = 60–80 nM for set **I**, depending on whether we use the mean Rhod-2 or mean Fluo-4 basal fluorescence values.

We now repeat the calculations of the previous section but using the values of P_0 prescribed by the simulations with $[Ca]_b$ = 100 nM for set **II** and $[Ca]_b$ = 50 nM for set **I** (Figure 2d). In this case, the r.h.s. estimates do not change much from the previous calculations. The l.h.s. estimates, on the other hand, change slightly coming closer together with the r.h.s. estimates. For example, using $V = 20$ μm/s, we obtain l.h.s.≈ 0.023 for $d = 0.6$ μm and l.h.s. ≈ 0.007 for $d = 1.4$ μm. If we use $V = 10$ μm/s, we obtain l.h.s.≈ 0.019 for $d = 0.6$ μm and l.h.s. ≈ 0.013 for $d = 1.4$ μm.

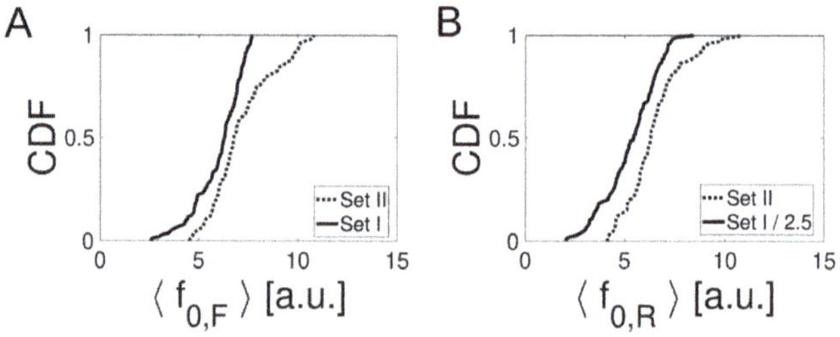

Figure 4. CDF of basal fluorescence. $\langle f_{0,D} \rangle$ for experiments with $[EGTA] = 90$ μM are shown in the Fluo-4 ($D = F$, (**a**)) and Rhod-2 ($D = R$, (**b**)) channels. In both cases, Set **II** is depicted with dotted lines and Set **I** with solid lines. In (**b**), the values, $\langle f_{0,R} \rangle$, of Set **I** are divided by $90/36 = 2.5$ to make both distributions comparable.

3. Discussion

Intracellular Ca^{2+} signals are ubiquitous across cell types. The spatial range over which they spread is key to determining the eventual end responses. This points to the importance of understanding how intracellular Ca^{2+} signals propagate inside the cells. To this end, Ca^{2+} release from the endoplasmic reticulum into the cytosol through IP_3Rs plays a major role. IP_3Rs are Ca^{2+} channels whose open probability depends on the cytosolic Ca^{2+} concentration. Therefore, the Ca^{2+} released through an open IP_3R can induce the opening of nearby IP_3Rs. High Ca^{2+} concentrations, on the other hand, lead to the inhibition of the channels. This dynamic is clearly excitable. In this regard, however, the excitability of the cytosol is "patchy": IP_3Rs tend to be organized in clusters separated by ~ 1.4 μm. This may lead to propagation failure when the Ca^{2+} released from one IP_3R reaches the vicinity of another one at a concentration that is not enough to induce its opening. The inter-cluster Ca^{2+}-mediated coupling can be interfered by means of Ca^{2+} buffers. This is used as an experimental tool, but the cells can do it as well.

In [10], we presented experimental results in which we studied how the presence of competing Ca^{2+} trapping mechanisms of different kinetics altered the resulting intracellular signals. Differently from previous studies [12], in [10], we made visible the invisible by using two dyes of different kinetics as the Ca^{2+} trapping mechanisms. The work of [10] not only allowed us to draw conclusions on how the signals were reshaped by the presence of the different buffers but also gave some indications on the spatial distribution of the IP_3Rs involved in the signals. In particular, based on a qualitative comparison between experiments and numerical simulations, we concluded in [10] that Ca^{2+} release seemed to occur not only from the clearly identifiable release sites (IP_3R clusters), but also from some functional, probably isolated, IP_3Rs in between them. In this paper, we have presented a quantitative analysis of the experiments of [10] that corroborated this conclusion.

For the quantitative comparison between experiments and models, in this paper, we have focused on the size distribution of the localized Ca^{2+} release events (puffs) that were obtained with the experiments of [10] for $[EGTA] = 90$ μM, $[F]_T = 36$ μM and two concentrations of the slow dye Rhod-2, $[R]_T = 36$ μM and 90 μM. Given that the fluorescence amplitude observed for a given release event could change with varying $[R]_T$ even if the underlying Ca^{2+} current remained the same, we characterized the observed puffs by the quantity, A_{lib-F} (Equation (11)) that we introduced in [10] to overcome this problem. The A_{lib-F} distributions obtained for the analyzed experiments showed a shift towards larger values of A_{lib-F} as $[R]_T$ was increased (see Figure 1). This shift agrees with previous observations and analyses according to which the increase of a slow Ca^{2+} buffer concentration (in this case, Rhod-2) disrupts the Ca^{2+}-mediated coupling between clusters (Figure 3a) [10,12,14,16]. Namely, we interpret this shift as reflecting the fact that events that are characterized by a certain number of simultaneously open IP_3Rs at a primary cluster and remain localized for a given value of

$[R]_T$ can induce the opening of IP_3Rs at other (secondary) clusters for smaller values of $[R]_T$. While the former events are puffs and would then be considered for the computation of the localized event size distribution, the latter would not because they correspond to waves.

We observed that clusters can become coupled when adding an IP_3R in-between them (Figure 3c). We introduced a probabilistic model in order to analyze quantitatively whether the differences observed for the experiments performed for $[R]_T = 0$ and $[R]_T = 90$ µM could be explained if Ca^{2+} release only occurred through IP_3R-clusters separated by 1.4 µm or not. Within the framework of the probabilistic model, we then combined the analysis of the experimental data with some probability estimates derived from numerical simulations similar to those presented in [10]. We determined in this way that the numerically estimated values were not compatible with the differences observed experimentally if the only Ca^{2+} release sites involved were $\sim$ 1.4 µm apart from one another. The experimental and numerical results were more compatible if we assumed that there was Ca^{2+} release from at least one IP_3R at a distance $\sim$ 0.6 µm from the primary Ca^{2+} release cluster. The presence of the slow dye, on the other hand, could reduce the basal Ca^{2+} concentration. We analyzed that possibility in the experimental data (Figure 4) and estimated that $[Ca]_b$ could have been reduced by half when $[R]_T$ was changed from 0 to 90 µM. The numerical simulations, on the other hand, showed that a decreasing value of $[Ca]_b$ with increasing $[R]_T$ gave better results for $d = 1.4$ µm (Figure 2d) in terms of their compatibility with their experiments. We then re-analyzed the experimental data but using numerical simulations that included this change in $[Ca]_b$ with varying $[R]_T$. The best situation to explain Figure 1 was obtained with simulations that combined a change in $[Ca]_b$ with $[R]_T$ and the presence of a functional IP_3R at a shorter distance ($d \sim 0.6$ µm) than the typical inter-cluster one.

Our quantitative analysis of the experiments of [10] presented in this paper confirms that the spatial landscape over which intracellular Ca^{2+} signals propagate do not consist solely of patches of excitability that are 1.4–2 µm apart from one another but that there are also "relay stations" (isolated functional IP_3Rs) in between. Probably, the existence of these in-between IP_3Rs is necessary for the propagation of Ca^{2+} waves.

4. Materials and Methods

4.1. Oocyte Preparation

Experiments were performed on *Xenopus laevis* immature oocytes previously treated with collagenase. Oocytes were loaded by intracellular microinjection with different compounds. Two calcium dyes Fluo-4 dextran high affinity ($K_d = 0.8$ µM) and Rhod-2 ($K_d = 2$ µM) were used to probe cytosolic [Ca]. Caged $InsP_3$ (D-Myo-Inositol 1,4,5-Triphosphate,P4(5)-(1-(2-Nitrophenyl)ethyl) Ester) was used to induce IP_3R opening. The exogenous Ca^{2+} buffer EGTA was also used. Final intracellular concentrations of the different compounds were calculated assuming a 1 µl cytosolic volume. Final intracellular concentration of $InsP_3$ was 9 µM in all of the experiments. The different concentrations used in each experiment are detailed in Table 1 where we classify the experiments in three sets. Fluo-4, Rhod-2 and $InsP_3$ were from Molecular Probes Inc.; EGTA was from Sigma Aldrich. Recordings were made at room temperature.

4.2. Confocal Microscopy

Confocal imaging was performed using a spectral confocal scanning microscope Olympus FluoView1000 that has a spectral scan unit connected to an inverted microscope IX81. The caged compound was photolyzed with the UV part of the spectrum of a mercury lamp that comes with the microscope using the modification introduced in [17]. Fluo-4 was excited with the 488 nm line of a multiline Argon laser, Rhod-2 was excited using the 543 nm line of a He–Ne laser. Both lasers were focused on the oocyte with a 60× oil immersion objective (NA 1.35). The Fluo-4 and Rhod-2 emitted fluorescences were simultaneously detected in the 500–600 nm and the 600–630 ranges, respectively, with PMT detectors. All the experiments were performed in the linescan imaging mode to improve the temporal resolution. Linescan images were obtained by scanning along a fixed line (250 px) within the oocyte. The acquisition rate was fixed at 10 μs per pixel resulting in a scan rate of 3.26 ms per line. The caged compound was photo-released approximately 3 s after the linescan acquisition started.

4.3. Image Analysis

All images were analyzed using routines written in MATLAB. In the experiments where we simultaneously acquired the fluorescence coming from two channels (around 510 nm for Fluo-4 and 570 nm for Rhod-2), we used a linear unmmixing method to minimize the effect of the spectral bleed-through ($R = 0.1626$ was the linear unmixing coefficient used). The images were also smoothed by averaging over the eight nearest pixels.

The events were identified and the images were processed as explained in [10]. From the fluorescence distribution, $f_D(x_i, t_j)$, collected in each of the channels ($D = R$ for Rhod-2 and $D = F$ for Fluo-4) at each pixel, (x_i, t_j), and the relative increase in fluorescence at the peak of the signal, $\Delta f_{r,D} = \max_{x_i,tj}((f_{r,D}(x_i, t_j) - f_{0,D}(x_i))/f_{0,D}(x_i))$, with $f_{0,D}(x_i)$ the mean basal fluorescence at x_i observed with D before the UV flash, we computed the corresponding Ca^{2+}-bound dye and (maximum) relative Ca^{2+} bound dye concentrations ($[CaD]$ and $\Delta[CaD]_r \equiv \max([CaD] - [CaD]_b)/[CaD]_b$, respectively, with $[CaD]_b$ the basal Ca^{2+}-bound dye concentration). To estimate the Ca^{2+}-bound dye concentration, we followed [18] neglecting fluctuations in the number of dye molecules that contribute to the fluorescence at each pixel, N_D, (for more details, see [10]):

$$[CaD] = \frac{[D]_T}{q_{1,D} - q_{2,D}} \left(\frac{f_D}{\gamma \langle N_D \rangle} - q_{2,D} \right), \quad D = R, F, \tag{9}$$

$$\Delta[CaD]_r = \Delta f_{r,D} \left(1 + \frac{q_{2,D}/q_{1,D}}{(1 - \frac{q_{2,D}}{q_{1,D}}) \frac{[Ca]_b}{[Ca]_b + K_{d,D}}} \right), \quad D = R, F. \tag{10}$$

To compute these quantities we followed [18] and used $\langle N_F \rangle = 32$, $\langle N_R \rangle = 32$ for $[R]_T = 36$ μM, $\langle N_R \rangle = 80$ for $[R]_T = 90$ μM, $[Ca]_b = 100$ nM, $q_{1,F} = 0.45$, $q_{2,F} = 0.01$, $K_{d,F} = 0.8$ μM, $q_{1,R} = 0.36$, $q_{2,R} = 0.02$ and $K_{d,R} = 2$ μM.

In this paper, we only analyzed the events observed in the Fluo-4 channel, i.e., for $D = F$. The event size of each analyzed puff was then characterized by the maximum value of the relative increase in the Ca^{2+}-bound Fluo-4 concentration, $\Delta[CaF]_r$ that we derived from the observed fluorescence. As done in [10], we then used the total Rhod-2 concentration, $[R]_T$, of the experiment to obtain estimates of the maximum values, $\Delta[CaF]_r([R] = 0)$, that would have been attained for the same release event if only the dye, Fluo-4, had been present. As discussed in [10], this estimate that we call A_{lib-F} is an increasing function of the Ca^{2+} current that underlies the release event regardless of the value of $[R]_T$ used if the Ca^{2+} current arises from a very localized spatial region (the cluster). As done in [10], we computed it as:

$$A_{lib-F} \approx \Delta[CaF]_r + \alpha_{F,R}[R]_T, \tag{11}$$

with $\alpha_{F,R} = 4.58 \times 10^{-3}$.

4.4. Basal Calcium Estimation

In order to study the behavior of the mean basal Ca^{2+} concentration, $[Ca]_b$, for each experiment type probed in the paper, we follow some of the steps of the method introduced in [18]. We work with linescan images obtained before any UV flash has been applied, i.e., we analyze basal fluorescence.

On these images, we get rid of the horizontal lines that are persistently dark, which correspond to the cortical granules. We then compute the mean basal fluorescence for each linescan image as:

$$\langle f_{0,D}\rangle = \frac{1}{N}\sum_{i\in bf}\sum_{j=1}^{juv} f_D(x_i,t_j), \quad D = R, F, \tag{12}$$

where the sum over i runs over the horizontal lines that are not persistently dark and the subscript, D, denotes whether the fluorescence comes from the Fluo-4 ($D = F$) or Rhod-2 ($D = R$) molecules. Using the values, $\langle f_{0,D}\rangle$, obtained for each experiment type, we compute the corresponding cumulative distribution functions of the mean basal fluorescence. To transform from basal fluorescence to basal Ca^{2+} , we use the following expression derived from [18]:

$$\overline{\langle f_{0,D}\rangle} = \gamma_D\left[(q_{1,D} - q_{2,D})\frac{[Ca]_b}{[Ca]_b + K_{d,D}} + q_{2,D}\right]\overline{\langle N_D\rangle}, \tag{13}$$

which takes into account the contributions to the fluorescence from the free and the Ca^{2+}-bound dye molecules with brightness $q_{2,D}$ and $q_{1,D}$, respectively. In Equation (13) $K_{d,D}$ is the dissociation constant of the Ca^{2+}-dye reaction, $\overline{\langle N_D\rangle}$ is the mean number of dye molecules that contribute to the fluorescence collected at a pixel and γ_D is a multiplying factor introduced by the detector ($\gamma_R = 6$ and $\gamma_F = 5$ [18]).

4.5. Numerical Simulations

To assess the rate of CICR-mediated coupling between neighboring clusters, we compute the probability that an IP_3R that n_s IP_3R located at a distance, d, from a Ca^{2+} point source becomes open during a time interval, Δt, since the start of the release by means of:

$$P_0(\Delta t, d, n_c, [R]_T) = 1 - \exp\left(-\int_0^{\Delta t} k_{on} n_s [Ca^{2+}](d,t)dt\right) \tag{14}$$

with $k_{on} = 20\ \mu M^{-1}s^{-1}$ the rate of Ca^{2+} binding to the activating site of an IP_3R of the DeYoung–Keizer model [19]. We compute $[Ca^{2+}](d,t)$ simulating a set of reaction-diffusion equations in a spherical volume (assuming spherical symmetry with r the radial coordinate) for: Ca^{2+}, an immobile endogenous buffer (S), two cytosolic indicators (F and R) and an exogenous mobile buffer ($EGTA$). A point source located at the origin and pumps (P) that remove Ca^{2+} uniformly in space are also included. For the source, we assume that it consists of n_c channels that open simultaneously at $t = 0$, each of which becomes close after a time that is drawn from an exponential distribution of mean $t_{open} = 20$ ms [19]. For the Ca^{2+}-buffer or dye reactions we consider that a single Ca^{2+} ion binds to a single buffer or dye molecule (X) according to:

$$Ca^{2+} + X \underset{k_{on-X}}{\overset{k_{off-X}}{\rightleftarrows}} [CaX], \tag{15}$$

where X represents F, R, $EGTA$, or S and k_{on-X} and k_{off-X} are the forward and backward binding rate constants of the corresponding reaction, respectively. We assume that the total concentrations of dyes and buffers ($[F]_T$,$[R]_T$, $[EGTA]_T$, and $[S]_T$) are spatially uniform at $t = 0$ so that they remain uniform and constant for all times. We also assume that $[Ca^{2+}]$ is initially uniform, equal to its basal value and in equilibrium with the buffers and dyes. The parameter values used are listed in Table 3.

Table 3. Parameter values used to solve the simulations.

Parameter	Value	Units
Free Calcium		
D_{Ca}	220	$\mu m^2 s^{-1}$
$[Ca]_b$	0.05–0.1	μM
Calcium dye Fluo-4-dextran		
D_F	15	$\mu m^2 s^{-1}$
k_{on-F}	240	$\mu M^{-1} s^{-1}$
k_{off-F}	180	s^{-1}
$[F]_T$	36	μM
Calcium dye Rhod-2-dextran		
D_R	15	$\mu m^2 s^{-1}$
k_{on-R}	70	$\mu M^{-1} s^{-1}$
k_{off-R}	130	s^{-1}
$[R]_T$	0, 36, 90	μM
Exogenous buffer EGTA		
D_{EGTA}	80	$\mu m^2 s^{-1}$
$k_{on-EGTA}$	5	$\mu M^{-1} s^{-1}$
$k_{off-EGTA}$	0,75	s^{-1}
$[D]_T$	90	μM
Endogenous immobile buffer		
D_S	0	$\mu m^2 s^{-1}$
k_{on-S}	400	$\mu M^{-1} s^{-1}$
k_{off-S}	800	s^{-1}
$[S]_T$	300	μM
Pump		
k_p	0.1	s^{-1}
v_p	0.9	$\mu M s^{-1}$
Source		
n_c	1, 10, 50	-
t_{open}	20	ms
I_{Ca}	0.1	pA

Author Contributions: E.P. conducted the experiments. E.P. and S.P.D. performed the simulations. E.P. and S.P.D. analyzed the experiments and simulations and wrote the paper. S.P.D. conceived the work.

Funding: This research has been supported by UBA (UBACyT 20020170100482BA) and ANPCyT (PICT 2015-3824).

Conflicts of Interest: The authors declare that the research was conducted in the absence of any commercial or financial relationships that could be construed as a potential conflict of interest.

References

1. Berridge, M.J.; Bootman, M.D.; Lipp, P. Calcium—A life and death signal. *Nature* **1998**, *395*, 645–648. [CrossRef] [PubMed]
2. Bootman, M.D.; Collins, T.J.; Peppiatt, C.M.; Prothero, L.S.; MacKenzie, L.; Smet, P.D.; Travers, M.; Tovey, S.C.; Seo, J.T.; Berridge, M.J.; et al. Calcium signalling—An overview. *Semin. Cell Dev. Biol.* **2001**, *12*, 3–10. [CrossRef] [PubMed]
3. Choe, C.U.; Ehrlich, B.E. The Inositol 1,4,5-Trisphosphate Receptor (IP3R) and Its Regulators: Sometimes Good and Sometimes Bad Teamwork. *Sci. Signal.* **2006**, *2006*, re15. [CrossRef] [PubMed]
4. Foskett, J.K.; White, C.; Cheung, K.H.; Mak, D.O.D. Inositol Trisphosphate Receptor Ca^{2+} Release Channels. *Physiol. Rev.* **2007**, *87*, 593–658. [CrossRef] [PubMed]
5. Fabiato, A. Calcium-induced release of calcium from the cardiac sarcoplasmic reticulum. *Am. J. Physiol.* **1983**, *245*, 1–15. [CrossRef] [PubMed]

6. Sun, X.P.; Callamara, N.; Marchant, J.S.; Parker, I. A continuum of InsP3-mediated elementary Ca^{2+} signalling events in Xenopus oocyte. *J. Physiol.* **1998**, *509*, 67–80. [CrossRef] [PubMed]
7. Smith, I.F.; Parker, I. Imaging the quantal substructure of single IP3R channel activity during Ca^{2+} puffs in intact mammalian cells. *Proc. Natl. Acad. Sci. USA* **2009**, *106*, 6404–6409. [CrossRef] [PubMed]
8. Solovey, G.; Dawson, S.P. Intra-cluster percolation of calcium signals. *PLoS ONE* **2010**, *5*, 1–8. [CrossRef] [PubMed]
9. Keizer, J.; Smith, G.D.; Ponce-Dawson, S.; Pearson, J.E. Saltatory Propagation of Ca^{2+} Waves by Ca^{2+} Sparks. *Biophys. J.* **1998**, *75*, 595–600. [CrossRef]
10. Piegari, E.; Lopez, L.F.; Dawson, S.P. Using two dyes to observe the competition of Ca^{2+} trapping mechanisms and their effect on intracellular Ca^{2+} signals. *Phys. Biol.* **2018**, *15*, 066006. [CrossRef] [PubMed]
11. Paredes, R.M.; Etzler, J.C.; Watts, L.T.; Zheng, W.; Lechleiter, J.D. Chemical calcium indicators. *Methods* **2008**, *46*, 143–151. [CrossRef] [PubMed]
12. Dargan, S.L.; Parker, I. Buffer kinetics shape the spatiotemporal patterns of IP3-evoked Ca^{2+} signals. *J. Physiol.* **2003**, *553*, 775–788. [CrossRef] [PubMed]
13. Dargan, S.L.; Schwaller, B.; Parker, I. Spatiotemporal patterning of IP3-mediated Ca^{2+} signals in Xenopus oocytes by Ca^{2+}-binding proteins. *J. Physiol.* **2004**, *556*, 447–461. [CrossRef] [PubMed]
14. Piegari, E.; Sigaut, L.; Ponce Dawson, S. Ca^{2+} images obtained in different experimental conditions shed light on the spatial distribution of IP3 receptors that underlie Ca^{2+} puffs. *Cell Calcium* **2015**, *57*, 109–119. [CrossRef] [PubMed]
15. Solovey, G.; Fraiman, D.; Dawson, S.P. Mean field strategies induce unrealistic nonlinearities in calcium puffs. *Front. Physiol.* **2011**, *2*, 1–11. [CrossRef]
16. Callamaras, N.; Parker, I. Phasic characteristic of elementary Ca^{2+} release sites underlies quantal responses to IP3. *EMBO J.* **2000**, *19*, 3608–3617. [CrossRef] [PubMed]
17. Sigaut, L.; Barella, M.; Espada, R.; Ponce, M.L.; Dawson, S.P. Custom-made modification of a commercial confocal microscope to photolyze caged compounds using the conventional illumination module and its application to the observation of Inositol 1,4,5-trisphosphate-mediated calcium signals. *J. Biomed. Opt.* **2011**, *16*, 066013. [CrossRef] [PubMed]
18. Piegari, E.; Lopez, L.; Perez Ipiña, E.; Ponce Dawson, S. Fluorescence fluctuations and equivalence classes of Ca^{2+} imaging experiments. *PLoS ONE* **2014**, *9*, e95860. [CrossRef] [PubMed]
19. Youngt, G.W.D.E.; Keizer, J. A single pool. *Nature* **1992**, *89*, 9895–9899.

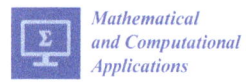

Mathematical and Computational Applications

Article

Structures and Instabilities in Reaction Fronts Separating Fluids of Different Densities

Johan Llamoza and Desiderio A. Vasquez *

Departamento de Ciencias, Sección Física, Pontificia Universidad Católica del Perú, Av. Universitaria 1801, San Miguel, Lima 32, Peru; jllamozar@pucp.pe

* Correspondence: dvasquez@pucp.edu.pe

Received: 22 April 2019; Accepted: 15 May 2019; Published: 17 May 2019

Abstract: Density gradients across reaction fronts propagating vertically can lead to Rayleigh–Taylor instabilities. Reaction fronts can also become unstable due to diffusive instabilities, regardless the presence of a mass density gradient. In this paper, we study the interaction between density driven convection and fronts with diffusive instabilities. We focus in fluids confined in Hele–Shaw cells or porous media, with the hydrodynamics modeled by Brinkman's equation. The time evolution of the front is described with a Kuramoto–Sivashinsky (KS) equation coupled to the fluid velocity. A linear stability analysis shows a transition to convection that depends on the density differences between reacted and unreacted fluids. A stabilizing density gradient can surpress the effects of diffusive instabilities. The two-dimensional numerical solutions of the nonlinear equations show an increase of speed due to convection. Brinkman's equation lead to the same results as Darcy's laws for narrow gap Hele–Shaw cells. For large gaps, modeling the hydrodynamics using Stokes' flow lead to the same results.

Keywords: reaction fronts; convection; diffusive instabilities

1. Introduction

Reaction fronts propagate in different media separating reactants and products. We find them in systems such as combustion [1,2], directional solidification [3], and waves of chemical activity [4]. In the latter case, a front develops due to the interaction between molecular diffusion and an autocatalytic chemical reaction [5]. Solutions of the reaction-diffusion equations correspond to fronts propagating in a given medium. For thin reaction fronts, these solutions can be approximated by an eikonal relation between the curvature and the normal component of the velocity [6]. The eikonal relation helped to explain the transition to convection for fronts in the iodate-arsenous relation, as well as the change of speed for propagating fronts in a Poiseuille flow [7,8] . Thin fronts showing diffusive instabilities can be modeled using the Kuramoto–Sivashinsky equation, which allows long wavelength instabilities for flat reaction fronts [9]. Experiments taken place in liquids require to take into account fluid motion as an additional component of the front dynamics. Fluid flow can be generated by mass density or surface tension gradients accross the front [10–12]. These convective flows will modify the shape of the front, and change its velocity [13,14].

Fronts propagating in the vertical direction can develop diffusive and Rayleigh–Taylor (RT) instabilities in liquids [15–17]. While diffusive instabilities are caused by differences in diffusivities, the Rayleigh–Taylor instability will take place if a fluid is placed under another fluid of larger density. Previous work using a reaction-diffusion model coupled to Darcy's law showed that diffusional instabilities enhance the RT instability, while buoyantly stable configurations can diminish the effects of diffusion-driven instabilities [15,16]. Using the KS equation coupled to Darcy's law showed the existence of stable cellular structures that involve convection [17]. For fronts propagating in viscous fluids, the KS equation coupled to the Navier–Stokes equation showed oscillatory instabilities depending on the viscosity reflected in a dimensionless Schmidt number [18].

In this paper, we explore the interaction between buoyancy driven flows and diffusive instabilities. Here differences in diffusivities may result in front instabilities even without fluid flow, making it a separate problem from double diffusion convection where buoyancy forces acting on different substances lead to fluid flow. The Kuramoto–Sivashinky equation will model diffusive instabilities in flat reaction fronts when coupled to fluid flow. We consider fluid flows inside Hele–Shaw cells or porous media described by Brinkman's equation. In a Hele–Shaw cell the fluid is confined between two vertical walls, Brinkmans equation considers the flow only in the direction parallel to the wall, thus becoming in a two-dimensional system. This approximation considers a new term that includes the dimensions of the gap. In the case of narrow gaps, Brinkmans equation can be approximated by Darcy's law, while for larger gaps the equations become the Navier–Stokes equations. We will carry out a linear stability analysis of the flat convectionless front for perturbations of fixed wavelength. We will also solve numerically the front evolution equations to obtain the patterns appearing in the nonlinear regime.

2. Equations of Motion

Reaction fronts that exhibit diffusive instabilities obey a system of reaction-diffusion equations that allow different diffusion coefficients for each substance. The resulting reaction fronts can be approximated by a front evolution equation, with the position of the front determined by a surface that separates reacted from unreacted species. In a two-dimensional cartesian coordinates (XZ) the front can be described by a height function $Z = H(X,t)$, with the time evolution of the height function H determined by a Kuramoto–Sivashinsky (KS) equation coupled to the fluid velocity $\vec{V}$:

$$\frac{\partial H}{\partial T} = V_0 + \mathcal{V}\frac{\partial^2 H}{\partial X^2} + \frac{V_0}{2}\left(\frac{\partial H}{\partial X}\right)^2 - \mathcal{K}\frac{\partial^4 H}{\partial X^4} + V_z|_H\,. \tag{1}$$

The parameters $\mathcal{V}$ and $\mathcal{K}$ depend on the diffusion coefficients for the different species. For zero fluid motion, there is a flat front solution that propagates with velocity V_0. The stability of this front is determined by the sign of the parameter $\mathcal{V}$. Small perturbations to the flat front of fixed wavelength can growth exponentially if the coefficient $\mathcal{V}$ is negative. This takes place if the wavelength is larger that a critical wavelength, perturbations of smaller wavelenghts will decay. In the opposite case ($\mathcal{V}$ positive), the front is stable regardless the wavelength of the perturbation. In the particular case of a system formed by two species with identical diffusivities involving cubic autocatalysis, the coefficients become $\mathcal{K} = 0$ and $\mathcal{V} = D$. In other systems, such as exothermic chemical reactions, the parameters $\mathcal{V} = 0$ and $\mathcal{K}$ will depend on the corresponding thermal diffusion coefficient [19,20]. The fluid flow appears as an addition to the normal front velocity of the propagating front [21]. The KS equation involves values up to second order in H, therefore near the onset of convection the normal component of the fluid flow corresponds to the addition of the vertical component of the fluid velocity V_z.

In this paper, we consider fluids inside Hele–Shaw cells or porous media, this flow can be described using Brinkman's equation:

$$\frac{\partial \vec{V}}{\partial T} + (\vec{V}\cdot\nabla)\vec{V} = -\frac{1}{\rho_0}\nabla P - 12\frac{\nu}{d^2}\vec{V} + \nu\nabla^2\vec{V} - \frac{\rho}{\rho_0}g\hat{z}\,. \tag{2}$$

In this equation P is the pressure, $\vec{V}$ is the fluid velocity, g is the acceleration of gravity, $\hat{z}$ is a unit vector pointing upward in the vertical Z-direction, ν is the coefficient of kinematic viscosity, ρ_0 is the density of the unreacted fluid, while ρ corresponds to the fluid density that depends on composition of the fluid. In a Hele–Shaw cell, the fluid is confined between two vertical walls separated by a gap width equal to d. For small gap widths, the fluid motion can be modeled with Darcy's law:

$$\vec{V} = -\frac{\kappa}{\rho_0 \nu}(\nabla P + \rho g \hat{z})P \tag{3}$$

Here κ is the coefficient of permeability for a porous media, which correponds to $d^2/12$ for flows in Hele–Shaw cells. Considering the density differences only in the large gravity term (Bousinesque approximation), the continuity equation is equal to

$$\vec{\nabla} \cdot \vec{V} = 0 \ . \tag{4}$$

The continuity equation in two-dimensions allows to derive the fluid velocity from a stream function $\Psi(X, Z, t)$ using the equations $V_x = \partial\Psi/\partial Z$ and $V_z = -\partial\Psi/\partial X$. With these relations we can eliminate the pressure term in Brinkman's equation to yield

$$\frac{\partial \Omega}{\partial t} = \frac{\partial(\Psi, \Omega)}{\partial(X, Z)} + \nu\nabla^2\Omega - 12\frac{\nu}{d^2}\Omega + \frac{g}{\rho_0}\frac{\partial \rho}{\partial X} \ . \tag{5}$$

The variable Ω in the last equation is the defined as the vorticity from

$$\Omega = \nabla^2\Psi \ . \tag{6}$$

We also define the following operator on two given functions F and G:

$$\frac{\partial(F, G)}{\partial(X, Z)} = \frac{\partial F}{\partial X}\frac{\partial G}{\partial Z} - \frac{\partial F}{\partial Z}\frac{\partial G}{\partial X} \tag{7}$$

The thin reaction front separates fluids of different densities, therefore fluid density can be written as

$$\rho = \rho_0 + \Delta\rho\Theta(Z - H) \ . \tag{8}$$

Here Θ corresponds to the theta function which takes the value of one if the argument is positive, and zero otherwise.

We can reduce the number of parameters under consideration defining appropriate dimensionless units. We introduce time and length scales defined by $L_T = \mathcal{K}/\mathcal{V}^2$, and $L_x = \sqrt{(\mathcal{K}/|\mathcal{V}|)}$. We define $|\mathcal{V}|$ as unit of the stream function, and $|\mathcal{V}|/L_x^2$ as unit of the vorticity. The variables in these units will be represented with lower case letters. The dynamic equations become

$$\frac{\partial \omega}{\partial t} = \frac{\partial(\psi, \omega)}{\partial(x, z)} + \mathrm{S_c}\nabla^2\omega - \alpha \mathrm{S_c}\omega - \mathrm{Ra}\mathrm{S_c}\frac{\partial h}{\partial x}\delta(z - h) \ . \tag{9}$$

This equation involves a dimensionless Rayleigh number

$$\mathrm{Ra} = \frac{g\delta\rho L^3}{\nu|\mathcal{V}|} \ , \tag{10}$$

a dimensionless Schmidt number

$$\mathrm{S_c} = \frac{\nu}{|\mathcal{V}|} \ , \tag{11}$$

and a parameter

$$\alpha = \frac{12\mathcal{K}}{d^2|\mathcal{V}|} \ . \tag{12}$$

In this system of units, the KS equation coupled to fluid flow becomes:

$$\frac{\partial h}{\partial t} = c_0 - \frac{\partial^2 h}{\partial x^2} + \frac{c_0}{2}\left(\frac{\partial h}{\partial x}\right)^2 - \frac{\partial^4 h}{\partial x^4} - \left.\frac{\partial \psi}{\partial x}\right|_h , \tag{13}$$

which involves a dimensionless front speed c_0 using the corresponding time and length scales. Reactions taking place in aqueous solutions have a large Schmidt number, therefore we will consider this limiting case where Equation (9) becomes

$$\nabla^2 \omega - \alpha\omega - \mathrm{Ra}\frac{\partial h}{\partial x}\delta(z-h) = 0 \, . \tag{14}$$

2.1. Linear Stability Analysis

The equations describing the evolution of the system (Equations (1) and (2)) allow a flat front solution moving with constant velocity c_0. We introduce perturbations of wavenumber q to this solution of the form

$$\psi = \hat{\psi}(z)e^{\sigma t}\sin(qx) \tag{15}$$

and

$$h = h_1 e^{\sigma t}\cos(qx) \tag{16}$$

leading to

$$(\frac{d^2}{dz^2} - q^2)^2\hat{\psi} - \alpha\omega \ + \mathrm{Ra}qh_1\delta(z) = 0 \tag{17}$$

and

$$\sigma h_1 = (q^2 - q^4)h_1 - q\hat{\psi}(0) \ . \tag{18}$$

The strategy for solving this system consists in first solving the linear equation Equation (17) in terms of h_1 and then substituting into Equation (18). The delta function leads to the following jump conditions at $z = 0$:

$$[\hat{\psi}] = [\frac{d\hat{\psi}}{dz}] = [\frac{d^2\hat{\psi}}{dz^2}] = 0 \ \text{ and } \ [\frac{d^3\hat{\psi}}{dz^3}] = -\mathrm{Ra}qh_1 \, . \tag{19}$$

$$(k^2 - q^2)(k^2 - \alpha - q^2) = 0 \ . \tag{20}$$

From here we obtain four values for k, namely $\pm q$, and $\pm k_\alpha$, where $k_\alpha = \sqrt{q^2 + \alpha}$. Considering that the stream function should vanish as the absolute value of the coordinate z becomes large, it lead us to write

$$\psi_1(z) = \begin{cases} Ae^{-qz} + Be^{-k_\alpha} & \text{if} z \geq 0 \\ Ce^{qz} + De^{k_\alpha} & \text{if } z < 0 \end{cases} , \tag{21}$$

Applying this for the stream function on the jump conditions across the front, we find a system of linear equations

$$A + B - C - D \quad = 0 \tag{22}$$

$$qA + k_\alpha B + qC + k_\alpha D \quad = 0 \tag{23}$$

$$q^2 A + k_\alpha^2 B - q^2 C - k_\alpha^2 D \quad = 0 \tag{24}$$

$$q^3 A + k_\alpha^3 B + q^3 C + k_\alpha^3 D \quad = \mathrm{Ra} q h_1 \tag{25}$$

The solution of this system determines the stream function given a particular value of h_1, therefore

$$\hat{\psi}(0) = A + B = -\frac{\mathrm{Ra} h_1 q}{2k_\alpha q(q + k_\alpha)}$$

Thus we obtain a dispersion relation between the growth rate σ and the wavenumber q which depends on the Rayleigh number and the parameter α which incorporates the gap width d on the model.

$$\sigma = q^2 - q^4 + \frac{\mathrm{Ra} q^2}{2k_\alpha q(q + k_\alpha)} \ . \tag{26}$$

2.2. Numerical Solutions

We use numerial methods to look for complex solutions of the nonlinear KS equation coupled to convective fluid motion. This solutions can be found in regimes where the flat front is unstable. Since the fluid flow is considered near the onset of convection, only linear terms for the velocity field are kept. To simplify further the problem allowing a direct comparison with Darcy's law limit, fluid boundary conditions are taken as slip-free boundaries. Therefore we can use a Fourier expansion on the front and stream function:

$$\psi = \sum_{n=1} \psi_n(z,t) \sin(nqx) \ , \tag{27}$$

and

$$h = \sum_{n=0} H_n(t) \cos(nqx) \ . \tag{28}$$

Here the domain varies from $x = 0$ to $x = L$ with the parameter q being equal to π/L. This expansion also incorporates the appropriate boundary conditions for h, which corresponds to vanishing first and third derivatives at the walls. Substituting into the linearized equation for the stream function allows to solve each Fourier coefficient in terms of the functions $H_n(t)$. This solution is similar to the one carried out in the linear stability analysis, therefore the KS equation with this solution becomes:

$$\frac{\partial h}{\partial t} = c_0 - \frac{\partial^2 h}{\partial x^2} + \frac{c_0}{2}\left(\frac{\partial h}{\partial x}\right)^2 - \frac{\partial^4 h}{\partial x^4} + \sum_n \frac{\mathrm{Ra} nq H_n}{2nqk_n(nq + k_n)} \tag{29}$$

The value of variable k_n is defined by $k_n = \sqrt{(nq)^2 + \alpha}$. Introducing the expansion on the nonlinear KS equation results in a set of coupled ordinary differential equations for the coefficients H_n

$$\frac{dH_0}{dt} = c_0 + c_0 \frac{q^2}{4} \sum_p p^2 H_p^2 \tag{30}$$

$$\begin{aligned} \tfrac{dH_n}{dt} &= [(nq)^2 - (nq)^4 + \tfrac{\mathrm{Ra}(nq)^2}{2k_n nq(nq + k_n)}] H_n + \\ & c_0 \tfrac{q^2}{4} \textstyle\sum_{l,p} l p H_p H_l \times (\delta_{n,|l-p|} - \delta_{n,l+p}) \ , \quad \text{for } n \geq 1 \ . \end{aligned} \tag{31}$$

In this work, we study the stability of convectionless flat front solutions (determined by Equation (26)), and steady curved fronts involving convection. To this end, we only consider the first 16 terms of Equation (31) setting the time derivatives equal to zero. The resulting nonlinear system was solved numerically using the scipy library from python through the module *optimize.fsolve*. Higher order truncations did not change the front solutions significantly. Each steady front solution has a constant velocity $c = dH_0/dt$ calculated using Equation (30). To address the issue of convergence, we displayed in Figure 1 the front speed relative to the flat front speed using an 8, 16, 24, and 32 terms truncation showing that the last three are indistinguishable from each other.

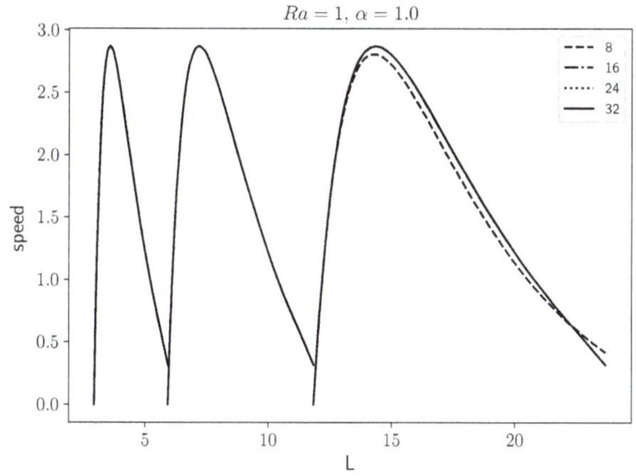

Figure 1. The front speed relative to the flat front speed as a function of domain size for series truncations of 8, 16, 24, and 32 terms. The last three lines are indistinguishable from each other.

We obtained solutions for different values of the parameters α, Ra, and L analyzing their stability, considering the flat front speed $c_0 = 1$. To determine the front stability, we used Python routines to obtain the eigenvalues of the Jacobian matrix derived from Equation (31). The sign of the largest real part of the eigenvalues determines the stability of the steady front, a negative sign will indicate stability. We also study solutions that evolve in time (such as oscillatory or chaotic solutions) using a simple Euler method to evolve Equation (31) with a time step $\Delta t = 10^{-3}$, and 18 term truncation.

3. Results

The density discontinuity accross the reaction front either enhances or inhibits the flat front instability found in the KS equation. Without fluid flow (Ra = 0), the dispersion relation Equation (26) has positive growth rates for perturbations of large wavelengths (small wavenumbers q) as shown in Figure 2. In this case perturbations of wavenumbers smaller than a critical value $q_c = 1$ have positive growth rates, indicating a flat front instability. In the case of positive Rayleigh numbers, where the less dense fluid is under a heavier fluid, the situation is similar. There is a critical wavenumber where perturbations with smaller wavenumber will grow, however this value is greater than one, the critical value without fluid flow. Therefore positive Rayleigh numbers will enhance the flat front instability. The opposite situation is found for negative Rayleigh numbers. Here the range of wavenumbers that allow negative growth rates diminishes. As we decrease the Rayleigh number from zero, the wavenumbers that lead to instabilities correspond to an interval that does not start in zero. That is near $q = 0$, the perturbations have negative growth rate, turning into positive growth rates as we increase q, and then becoming negative once again as we increase q further. This interval diminishes, and finally disappears, as we decrease the Rayleigh number towards negative values. Therefore, for a given negative value of the Rayleigh number the flat front becomes stable. A large enough density gradient is able to stabilize the flat fronts described by the KS equation.

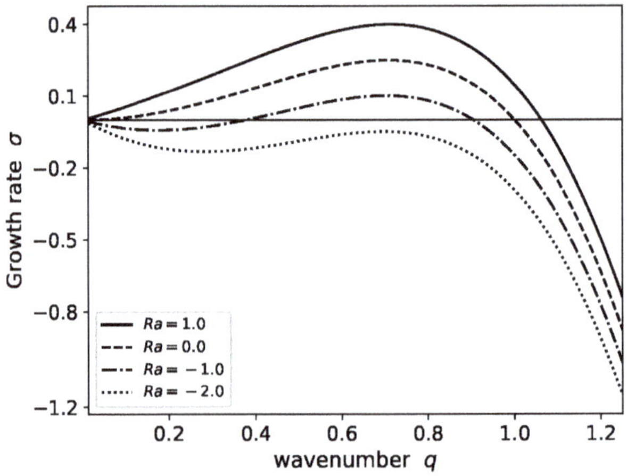

Figure 2. Growth rate for perturbations of wavenumber q. Negative Rayleigh numbers lead to negative growth rates near $q = 0$. For Ra $= -2$ the flat front is stable.

Perturbations of fixed wavenumber can be stabilized with an appropriate Rayleigh number. Figure 3 shows the Rayleigh number necessary for growth rate equal to zero for different values of the wavenumber q.

In this figure, we compare three hydrodynamic models including Brinkman's equation. The curve for the Brinkman's model has a minimum, therefore a stable flat front requires a Rayleigh number below the minimum. This curve also shows that perturbations with wavenumber below 1 require a negative Rayleigh number to avoid a growing perturbation. For Rayleigh numbers that are negative but above the minimum, there is an interval in the values of q with positive growth rates. In these cases the front can be stable for long wavelength perturbations (small values of q). In Figure 3 we also display the results using two other hydrodynamic models: Darcy's law and the Stokes equation. In a Hele–Shaw cell width a narrow gap (as is the case for $\alpha = 40$), the results of Darcy's law are close to the results using Brinkman's equation. Flows in Hele–Shaw cells with narrow gaps can be described with Darcy's law, which is the limiting case of Brinkman's equation. With a wider gap ($\alpha = 0.1$), we find that the results using the Stokes equation are closer to the results of Brinkman's equation (Figure 4), approaching the correct wide gap limit. Brinkman's equation provide us the results between the two limits.

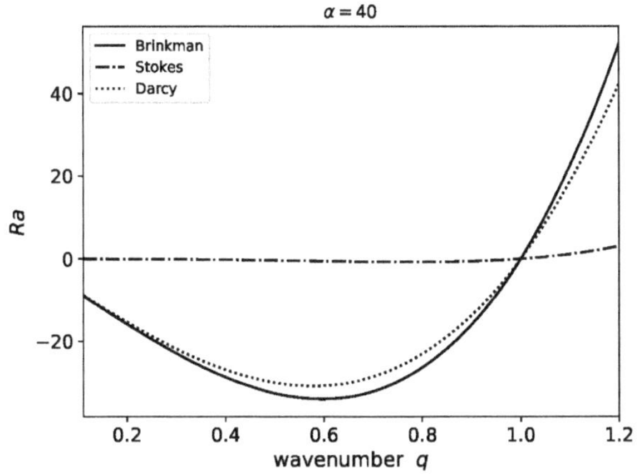

Figure 3. Rayleigh number necessary for zero growth rate as a function of wavenumber q. For a given wavenumber, higher Rayleigh numbers will result in growing perturbations. We compare the results for three hydrodynamic models. In the case of small gap ($\alpha = 40$) results using Brinkman's equation are close to results using Darcy's law.

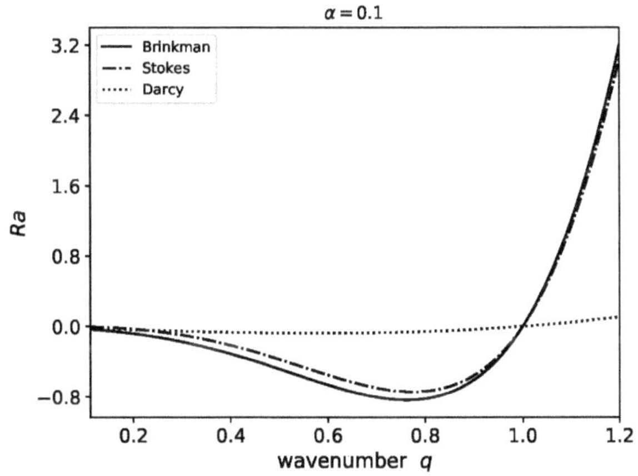

Figure 4. Rayleigh number necessary for zero growth rate as a function of wavenumber q. For a given wavenumber, higher Rayleigh numbers will result in growing perturbations. We compare the results for three hydrodynamic models. In the case of a large gap ($\alpha = 0.2$) results using Brinkman's equation are close to results using Stokes equation.

Flat fronts propagating in two-dimensional rectangular domains (resembling vertical tubes), can be stable depending on the value of the Rayleigh number. Figure 5 displays the largest growth rate as a function of Rayleigh number and domain width. A solid curve separates the regions of positive and negative growth rate, consequently values under the curve indicate a stable flat front. Narrow rectangular domains of width equal to L allow only perturbations of wavenumber greater than $q = \pi/L$. Since perturbations of $q < 1$ require positive Rayleigh numbers to generate instabilities, domains with $L < \pi$ will require a less dense fluid is under a more dense fluid to trigger an instability. On the contrary, if $L > \pi$, the flat front requires a negative Rayleigh number to be stable. We observe in Figure 5 that the critical Rayleigh number for front instability fluctuates as a function of the domain width. This implies that a flat front can be unstable for a certain width, but increasing the width could stabilize the flat front, if the original Rayleigh number is now below the new critical value.

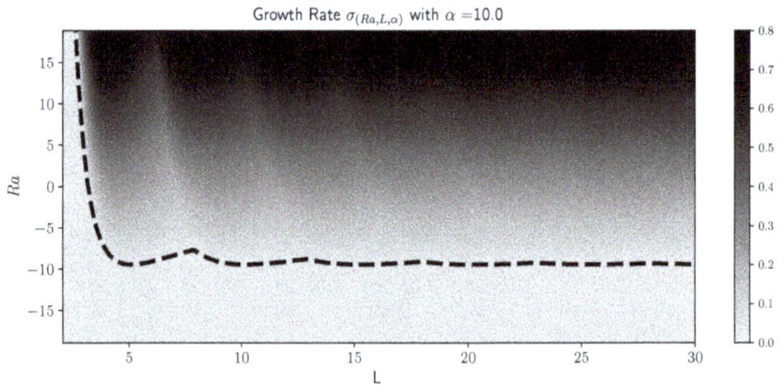

Figure 5. Largest growth rate as a function of Rayleigh number and domain width. The broken line indicates the value of growth rate zero.

Increasing the domain size beyond the critical value for the flat front instability leads to convective fluid motion. Numerical solutions of the nonlinear equations just above criticality show that small random perturbations to the flat front grow with time. After some time, they form a steady pattern with fluid rising on one side of the domain and falling on the opposite side generating a single convective roll as shown in Figure 6a. The mirrored solution is also a steady solution of the equation, that developes from different initial conditions, and is also stable. The patterns propagate with constant shape and a velocity higher than the velocity for the flat front. This shape also leads to steady front solutions for rectangular domains of larger widths. For example, doubling the domain size we can accomodate two single counterrotating convective rolls as shown in Figure 6b. These solutions have the same speed as the single roll solution, but in this case the fluid rises at the center of the rectangular domain falling on the sides. The resulting shape corresponds to a symmetric front with a maximum near the center of the domain. We can continue this process for larger domains, finding structures formed by placing a one roll solution, next to another counterrotating one roll solution. The first structures formed in this manner are displayed in Figure 6. The solutions with an even number of convective rolls are symmetric with respect to the center of the domain. The mirrored solution is also a solution, for the case of fronts with odd number of rolls. This solution resemble the cellular solutions found in the Kuramoto–Sivashinky equation [20]. We will analyze later the conditions for stabily of the cellular solutions.

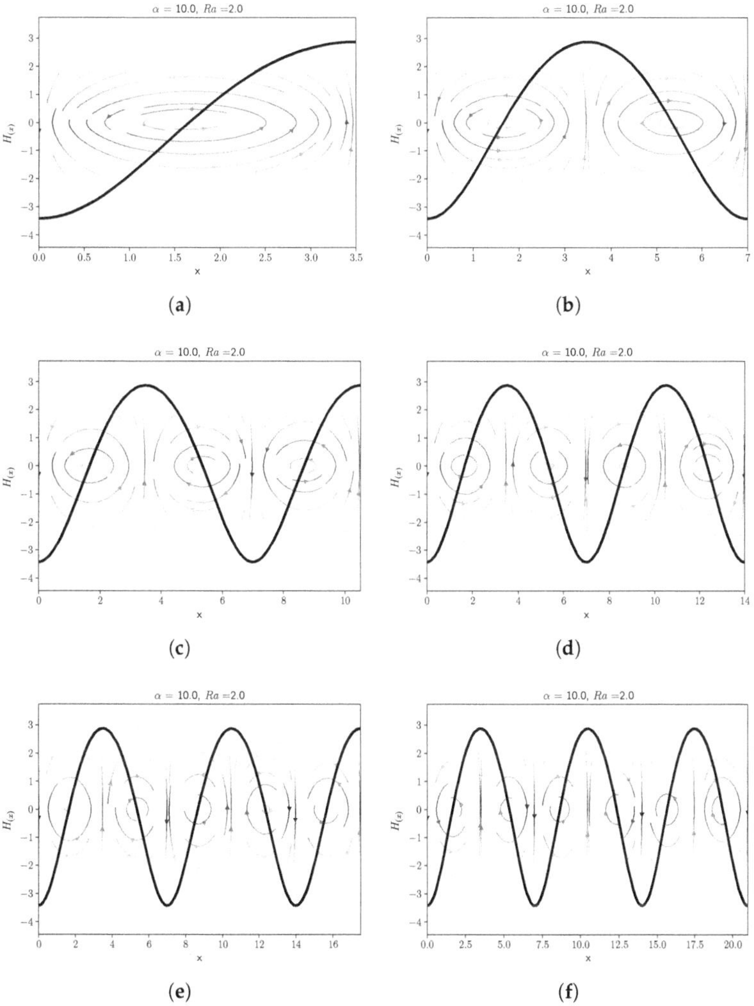

Figure 6. Steady cellular structures. Each structure is build by using a one roll solution (**a**) next to another one roll solution. Here we show structures with (**b**) two, (**c**) three, (**d**) four, (**e**) five, and (**f**) six convective rolls.

Convection increases the speed of propagating fronts as they propagate upward in narrow vertical rectangles. Progating fronts of steady shape develop when the width of the tube is larger than a critical width. In Figure 7a we display the speed of fronts relative to the flat front speed ($c - c_0$) as a function of the domain width for a positive Rayleigh number (Ra $= 1$). Below the critical width, the only solution corresponds to the flat front solution. As we increase the width above the critical width the solution with one convective roll appears. The analytical linear stability analysis shows that the flat front is unstable beyond the critical width. The linear stability analysis of the curved fronts were carried out obtaining the Jacobian matrix on the steady solutions of Equation (31) as described in Section 2.2.

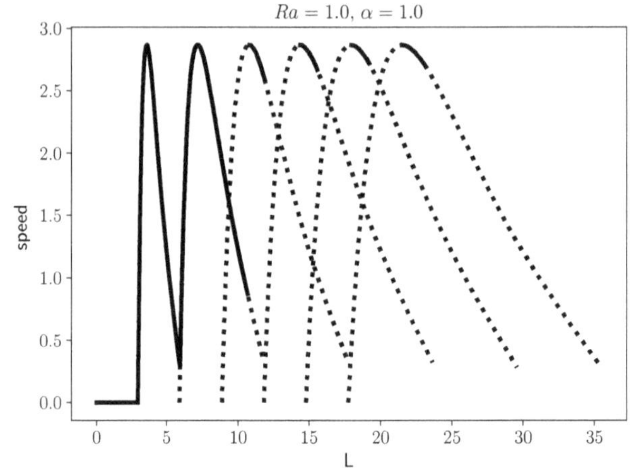

(**a**)

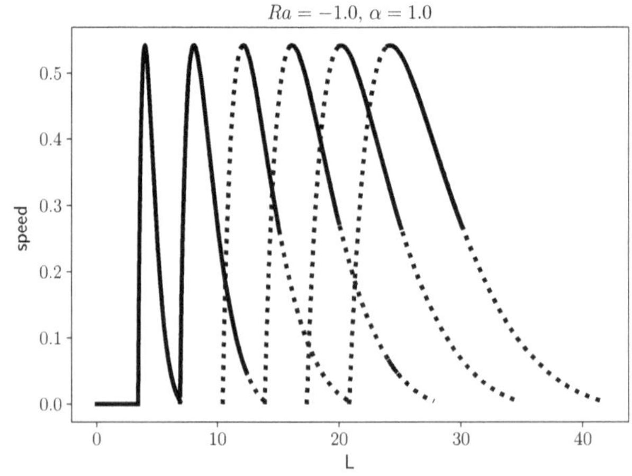

(**b**)

Figure 7. Increase of speed for different steady cellular structures as a function of domain width *L*. The solid lines indicate that the structure is stable. A broken line corresponds to unstable structures. (**a**) Corresponds to Ra = 1, while (**b**) corresponds to Ra = −1.

This linear stability analysis determines that the single roll solution is stable for widths above, but close, to the critical width. Increasing the width further also increases the speed of the convective front until it reaches a maximum value. After the maximum value is reached, the front speed decreases until the one roll solution is not available, at this point there is another solution containing two counterrotating convective rolls. The linear stability analysis shows that the one-roll solution is always stable. There is a small range of values that allows both the one-roll, and the two-roll solution, in this case one the one-roll solution is stable. Once the one-roll solution is no longer available, the two-roll solution appears, increasing the speed as the width is increased, reaching a maximum, and finally disappearing. However, contrary to the one-roll solution, the two roll solution is not always stable, there are regions of instability. Figure 7 also displays the speed relative to the flat front speed of steady solutions with three, four, and five rolls. In all these cases, the behavior is similar, the speed increases, reaches a maximum, after that decreases, and then the solution is no longer available. However in all these cases, the solution is unstable for most values, only for small ranges the solution is stable. For negative values of the Rayleigh number (Ra < 0), the critical width for convective fronts decreases (Figure 7b). The speed of the cellular structures have a similiar behavior as for positive Rayleigh numbers, that is they show width values where the solution exist, achieving a maximum speed for certain widths. However, in this case the speeds are lower, but the width values showing stable fronts is larger. Having the less dense fluid above the more dense fluid contributes to stabilize structures, but with a corresponding decrease in velocity. Lowering the Rayleigh number to Ra $= -1.6$ we find even lower convective front velocities relative to the flat front speed (Figure 8). In this case increasing the width leads to a region where the convective solution is no longer possible, where the flat front is stable once again. Increasing the width will lead to effectively stopping convection.

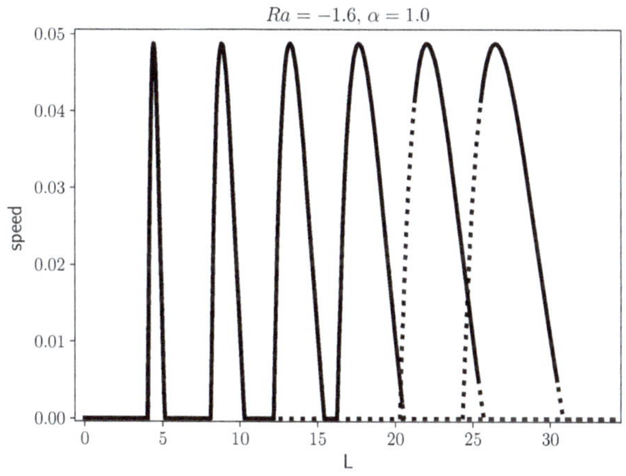

Figure 8. Relative speed (speed minus the flat front speed) for different steady cellular structures as a function of domain width L. The solid lines indicate that the structure is stable. A broken line corresponds to unstable structures. For Ra $= -1.6$ we find widths with a stable cellular structure, but increasing the width only allows stable flat fronts.

Fronts described by the KS equation exhibit complex spatio-temporal behavior, such as oscillations and chaos. Fluid motion and confinement can help to enhance or surpress this behavior. In Figure 9 we display a bifurcation diagram showing the relative maximum front speed relative to the flat front speed as a function of the Rayleigh number. As the front evolves with time its propagating velocity changes, reaching a maximum before slowing down. Oscillatory fronts will display a few maxima before repeating the sequence again. For values of the Rayleigh number close to Ra $= 0.1$, Figure 9 displays three velocity maxima. As the Rayleigh number increases, we observe a sequence of bifurcations to oscillatory states with higher periodicity. This period doubling cascade leading to chaotic states is the result of enhanced fluid motion at higher Rayleigh numbers. Negative Rayleigh numbers inhibit the oscillatory motion resulting in flat front thus surpressing complex fluid motion. Another parameter that has an impact on complex behavior is the parameter α, which is inversely proportional to the square of the gap width. Figure 10 shows a similar period doubling cascade displaying the front velocity maxima relative to the flat front speed versus the parameter α. We notice that for α close to 1.34 there is clearly periodic behavior. Reducing α increases the number of maxima, eventually reaching a chaotic state. This implies that confining the substances in a Hele–Shaw cell of small gap (large α) diminish the complex spatio-temporal behavior. This implies that in experiments in a Hele–Shaw cell, the gap width can be an effective control parameter for transitions between different spatio-temporal states.

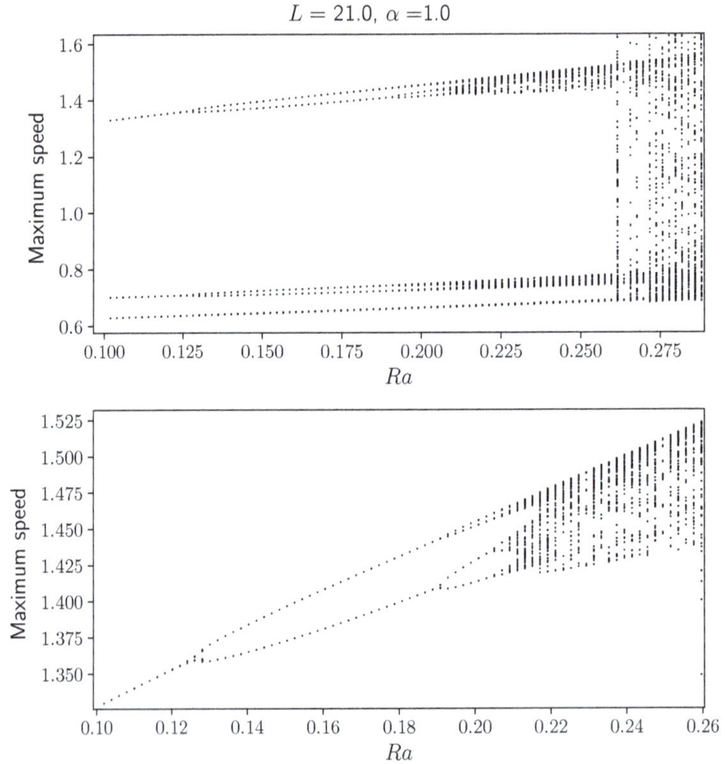

Figure 9. Bifurcation diagrams showing relative maxima for the velocity of the average front position relative to the flat front speed. We observe a period doubling cascade to chaos as the Rayleigh number is increased. The bottom panel is a detailed from the top panel.

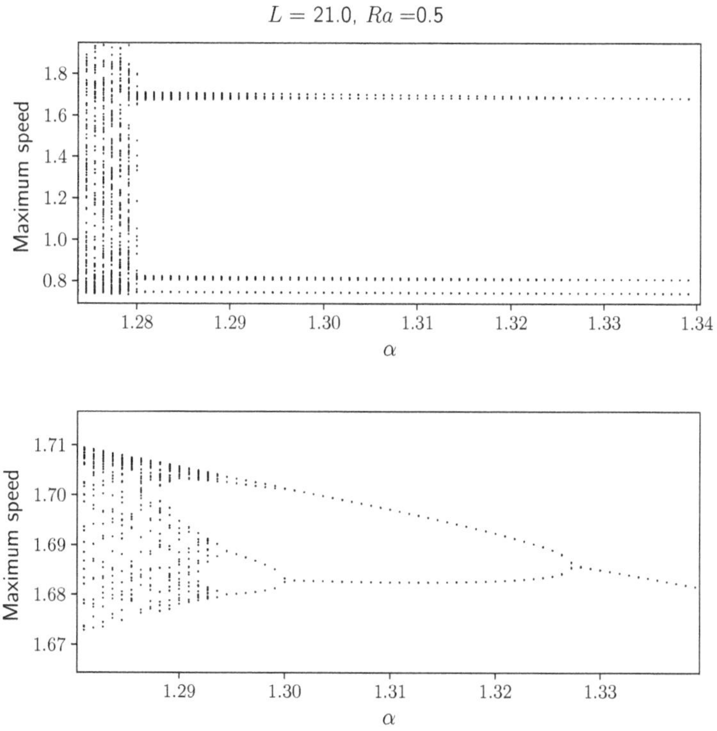

Figure 10. Bifurcation diagrams showing relative maximums for the velocity of the average front position relative to the flat front speed. We observe a period doubling cascade to chaos as the parameter α is decreased. This corresponds to reducing the gap between the plates in a Hele–Shaw cell. The bottom panel is a detailed from the top panel.

4. Summary and Discussion

We found that the presence of density gradients in fronts governed by the KS equation could enhance or supress complex behavior. Positive density gradients (Ra > 0) lowers the critical wavelength for onset of convection, thus fronts propagating in rectangular domains would require smaller widths for flat front stability. On the contrary, negative Rayleigh numbers provide a mechanism to diminish the instability found in the KS equation. Fronts propagating with parameters just above the flat front instability evolve into a curved front of constant shape involving a single convective roll. Placing this single roll solution (or cell) next to each other results in cellular structures that are extended solutions on a larger domain. We analyzed the stability of these structures as a function of the domain widths finding corresponding regions of stability. A density gradient favorable to convection diminishes the widths for stability, on the other hand, a gradient unfavorable to convection increases the stability of cellular structures. Decreasing the negative Rayleigh number even further leads to a situation where unstable flat fronts can be stabilized by increasing the width of the domain. We also found that complex behavior can be surpressed by either reducing the Rayleigh number, or by controlling the gap width in Hele–Shaw cells. The latter being a mechanism that can useful to experiments. A connection with experiments will require knowledge of the parameters $\mathcal{V}$ and $\mathcal{K}$. An estimate can be obtained by using the wavelength for maximum growth rate for purely diffusive instabilities, which in dimensioned units corresponds to $\lambda = 2\pi/q = 2\pi\sqrt{2\mathcal{K}/\mathcal{V}}$. A second relation between $\mathcal{V}$ and $\mathcal{V}$ can be obtained from the experimental front speed V_0, which in dimensionlesss units correspond to $c_0 = V_0 L_T / L_x = 1$. Using these relations as a guide, we can compute the value of the Rayleigh number using $\mathrm{Ra} = g\delta\lambda^2/8\pi^2\nu c_0$. We can use a typical wavelength of about 1 cm observed in experiments [22] together with parameters from chemical reaction fronts propagating in aqueous solutions [7] obtaining a Rayleigh number of -45, where the more dense fluid is above the fluid of smaller density. This Rayleigh number requires a value of $\alpha = 53.8$ to inhibit the diffusive instability, which leads to a distance between plates in a Hele–Shaw cell equal to 0.5 mm. This distance is small, but larger gaps can be used if the absolute value of the Rayleigh number can be diminished. Therefore experiments can be carried out for different gap values of Hele–Shaw cells to test how the density differences between reacted and unreacted fluids can supress the diffusive instability. One of the advantage of using Brinkman's equation in the calculations is that it leads to the correct narrow and wide gap limits, which are governed by Darcy's law and the Navier–Stokes equations.

Author Contributions: Conceptualization, D.A.V. and J.L.; methodology, D.A.V. and J.L.; software, D.A.V. and J.L.; validation, D.A.V. and J.L.; investigation, D.A.V. and J.L; writing–original draft preparation, D.A.V.; writing–review and editing, D.A.V. and J.L.; visualization, J.L.

Funding: J.L. received funding through a sholarship from the Consejo Nacional de Ciencia, Tecnología e Innovación Tecnológica under grant 236-2015-CONCYTEC.

Conflicts of Interest: The authors declare no conflict of interest.

References

1. Vladimirova, N.; Rosner, R. Model flames in the Boussinesq limit: The effects of feedback. *Phys. Rev. E* **2003**, *67*, 066305. [CrossRef] [PubMed]
2. Rakib, Z.; Sivashinsky, G.I. Instabilities in upward propagating flames. *Combust. Sci. Technol.* **1987**, *54*, 69–84. [CrossRef]
3. Langer, J.S. Instabilities and pattern formation in crystal growth. *Rev. Mod. Phys.* **1980**, *52*, 1–28. [CrossRef]
4. Winfree, A.T. Spiral waves of chemical activity. *Science* **1972**, *175*, 634–636. [CrossRef]
5. Showalter, K. Quadratic and cubic reaction-diffusion fronts. *Nonlinear Sci. Today* **1995**, *4*, 1–2.
6. Tyson, J.J.; Keener, J.P. Singular perturbation theory of traveling waves in excitable media (a review). *Phys. D Nonlinear Phenom.* **1988**, *32*, 327–361. [CrossRef]
7. Edwards, B.F.; Wilder, J.W.; Showalter, K. Onset of convection for autocatalytic reaction fronts: Laterally unbounded system. *Phys. Rev. A* **1991**, *43*, 749–760. [CrossRef] [PubMed]
8. Spangler, R.S.; Edwards, B.F. Poiseuille advection of chemical reaction fronts: Eikonal approximation. *J. Chem. Phys.* **2003**, *118*, 5911–5915. [CrossRef]
9. Malevanets, A.; Careta, A.; Kapral, R. Biscale chaos in propagating fronts. *Phys. Rev. E* **1995**, *52*, 4724–4735. [CrossRef]
10. Showalter, K.; Vasquez, D.A.; Masere, J.; Wilder, J.W.; Edwards, B.F. Nonaxisymmetric and axisymmetric convection in propagating reaction-diffusion fronts. *J. Phys. Chem.* **2005**, *98*, 6505–6508. [CrossRef]
11. Böckmann, M.; Rakotomalala, N.; Salin, D.; Böckmann, M. Buoyancy-driven instability of an autocatalytic reaction front in a Hele–Shaw cell. *Phys. Rev. E* **2002**, *65*, 051605. [CrossRef]
12. Bába, P.; Rongy, L.; de Wit, A.; Hauser, M.J.B.; Tóth, Á.; Horváth, D. Interaction of pure Marangoni convection with a propagating reactive interface under microgravity. *Phys. Rev. Lett.* **2018**, *121*, 024501. [CrossRef]
13. Vasquez, D.A.; Littley, J.M.; Wilder, J.W.; Edwards, B.F. Convection in chemical waves. *Phys. Rev. E* **1994**, *50*, 280–284. [CrossRef]
14. Elliott, D.; Vasquez, D.A. Convection in stable and unstable fronts. *Phys. Rev. E* **2012**, *85*, 1–6. [CrossRef]

15. D'Hernoncourt, J.; Merkin, J.H.; de Wit, A. Interaction between buoyancy and diffusion-driven instabilities of propagating autocatalytic reaction fronts. I. Linear stability analysis. *J. Chem. Phys.* **2009**, *130*, 114502. [CrossRef]
16. D'Hernoncourt, J.; Merkin, J.H.; de Wit, A. Interaction between buoyancy and diffusion-driven instabilities of propagating autocatalytic reaction fronts. II. Nonlinear simulations. *J. Chem. Phys.* **2009**, *130*, 114503. [CrossRef] [PubMed]
17. Vilela, P.M.; Vasquez, D.A. Rayleigh–Taylor instability of steady fronts described by the Kuramoto–Sivashinsky equation. *Chaos* **2014**, *24*, 023135. [CrossRef]
18. Coroian, D.; Vasquez, D.A. Oscillatory instability in a reaction front separating fluids of different densities. *Phys. Rev. E* **2018**, *98*, 1–8. [CrossRef]
19. Michelson, D.M.; Sivashinsky, G.I. Nonlinear analysis of hydrodynamic instability in laminar llames—II. Numerical experiments. *Acta Astronaut.* **1977**, *4*, 1207–1221. [CrossRef]
20. Sivashinsky, G.I. Instabilities, pattern formation, and turbulence in flames. *Annu. Rev. Fluid Mech.* **1983**, *15*, 179–199. [CrossRef]
21. Wilder J.W.; Vasquez, D.A.; Edwards, B.F. Modification of the eikonal relation for chemical waves to include fluid flow. *Phys. Rev. E* **1998**, *47*, 3761–3764. [CrossRef]
22. Horváth, D.; Tóth, Á. Diffusion-driven front instabilities in the chlorite-tetrationate reaction. *J. Chem. Phys.* **1998**, *108*, 1447–1451. [CrossRef]

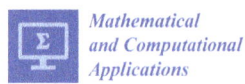

Mathematical and Computational Applications

MDPI

Article

Exact Evaluation of Statistical Moments in Superradiant Emission

Gilberto M. Nakamura [1], Brenno Cabella [2] and Alexandre S. Martinez [2,3,*]

[1] Faculdade de Filosofia, Ciências e Letras de Ribeirão Preto (FFCLRP), Universidade de São Paulo (USP), Avenida Bandeirantes 3900, 14040-901 Ribeirão Preto, São Paulo, Brazil; gmnakamura@usp.br

[2] Instituto de Física Teórica, Universidade Estadual Paulista (UNESP), Rua Dr. Bento Teobaldo Ferraz 271, 01140-070 São Paulo, Brazil; cabellab@gmail.com

[3] Instituto Nacional de Ciência e Technologia de Sistemas Complexos (INCT-SC), Rua Dr. Xavier Sigaud 150, Urca, 22290-180 Rio de Janeiro, Brazil

* Correspondence: asmartinez@usp.br

Received: 30 April 2019; Accepted: 19 June 2019; Published: 23 June 2019

Abstract: Superradiance describes the coherent collective radiation caused by the interaction between many emitters, mediated by a shared electromagnetic field. Recent experiments involving Bose–Einstein condensates coupled to high-finesse cavities and interacting quantum dots in condensed-matter have attracted attention to the superradiant regime as a fundamental step to create quantum technologies. Here, we consider a simplified description of superradiance that allows the evaluation of statistical moments. A correspondence with the classical birthday problem recovers the statistical moments for discrete time and an arbitrary number of emitters. In addition, the correspondence provides a way to calculate the degeneracy of the problem.

Keywords: stochastic processes; complex systems; self-organization; Dicke model; birthday problem

1. Introduction

Superradiance describes the coherent collective radiation caused by the interaction between N emitters, mediated by a shared electromagnetic field [1]. The phenomenon occurs when the average distance between emitters is smaller than the emitted wavelength, producing an emission pattern that differs markedly from the spontaneous emission of photons by isolated atoms. Supporting experimental evidence shows that the electric dipole of atoms in the atomic cloud becomes strongly correlated, producing coherent radiation less sensitive to thermal fluctuations of optical cavities [2–4], which can be used to improve the precision of atomic clocks [5]. Dicke first predicted superradiance after considerations on symmetries and conservation laws of total angular momentum in matter–light interaction in cavities. Almost 70 years after its discovery, superradiance remains a topic of intense experimental and theoretical research in quantum many-body physics [6,7].

Although its origins can be tracked down to the foundation of quantum optics, superradiance has found applications in condensed-matter physics. More specifically, the experimental observation of superradiance in ensembles of quantum dots suggests long-range electromagnetic interactions can be fine-tuned for creating quantum technologies, by selecting an appropriate density of quantum dots, their physical dimensions, and their atomic composition or the surrounding media [8]. The iconic spontaneous mission of photons by emitters may not occur as N independent events, even for $N = 2$, as verified in recent experiments [9]. Enhanced spontaneous emission can also occur due to strong correlations between quantum dots immersed in confining potentials [10]. More broadly, theoretical and experimental studies of superradiance in condensed-matter have led to reexamination of collective effects in nanoscale systems [11,12], metallic nanoparticles and nanostructures [13,14], and magnetic nanosystems, as well [15–18].

Meanwhile, experimental realizations of the superradiant regime in Bose–Einstein condensates coupled with optical cavities have provided insights into quantum phase transition between normal and superradiant phases [19,20]. In these experiments, the atomic condensate remains trapped inside a high finesse optical cavity. A continuous wave with the frequency far detuned from relevant atomic transition is pumped onto the condensate, transverse to the axis of the optical cavity. Because the pump wave is detuned, the interaction is dispersive with negligible spontaneous emission. As a result, waves are scattered into the cavity with resonant frequency ω, which self-organize the condensate in a spatial lattice, with half-wavelength spacing. Even more, the resulting lattice acquires one out of two possible orientations, suggesting a $\mathbb{Z}_2$ spontaneous symmetry-breaking at the quantum phase transition [21]. Indeed, a phase transition in the Dicke model has been predicted by Hepp and Lieb [22]. The Hamiltonian of the Dicke model reads:

$$\hat{H} = \hbar\omega\, \hat{a}^\dagger \hat{a} + \hbar\omega_0\, \hat{J}_z + \frac{\gamma}{\sqrt{N}} \left(\hat{a}^\dagger + \hat{a}\right) \hat{J}_x, \tag{1}$$

formed by N two-level emitters with energy $\hbar\omega_0$ and a single bosonic mode with cavity frequency ω and coupling γ. The collective pseudospin operators $\hat{J}_\mu$ ($\mu = x, y, z$) satisfy the usual relations $[\hat{J}_\alpha, \hat{J}_\beta] = i\hbar \sum_\mu \varepsilon_{\alpha\beta\mu} \hat{J}_\mu$. The light–matter interaction is described by the operator $(\hat{a}^\dagger + \hat{a})\hat{J}_x = (1/2)(\hat{a}^\dagger \hat{J}_- + \hat{a}\hat{J}_-) + (1/2)(\hat{a}^\dagger \hat{J}_+ + \hat{a}\hat{J}_-)$. This operator carries two familiar transitions, namely the emission of a photon to the cavity and the photon absorption, as well as two apparently non-conservative energy transitions: photon emission followed by atomic excitation and photon absorption followed by atomic relaxation. They appear non-conservative because pump waves are not taken into account, resulting in an open system. Finally, the intensity I produced by a superradiant system contains contributions proportional to $|\langle j, m_z \pm 1|\hat{J}_\pm|j, m_z\rangle|^2 = j(j+1) - m(m \pm 1)$. For $j = N/2$, transitions starting from quantum states with $m_z = 0$ maximize the emission of radiation and produce $I \propto N^2$, which is one of the hallmarks of superradiance. The intensity diminishes for other values of $|m_z|$, becoming a linear function of N for $|m_z| = j = N/2$.

To date, Bose–Einstein experiments have offered the most flexibility to select couplings and explore the superradiant phase transition and its critical properties in detail. Among them is the emergence of long-range correlations among emitters, which can be further explored to create faster procedures for information storage and retrieval in qubit networks [23]. The key aspect to understand the underlying physics in the Dicke model is the proper comprehension of self-organizing phenomena [12]. The ordered collective behavior arises from small fluctuations and develops positive feedback [24–29]. These collective effects are observable and reveal the effect of higher order statistical moments and correlation functions. In the Dicke model, several approaches have been proposed to capture the minimal cooperative properties. In the thermodynamic limit ($N \gg 1$), the model has been studied in the rotating wave approximation (RWA) [22,30]. More recently [31], without RWA and using the Holstein–Primakoff transformations with $N \gg 1$, Emary and Brandes presented an exact solution, where they verified the existence of quantum phase transitions and the emergence of a chaotic regime. For finite N, the interference among confined emitters is enhanced by finite size effects, and the model is non-integrable [32], while the available solutions are restricted to numerical ones [33,34].

More recently, semiclassical approximations have provided a far more concrete structure of the density of states of the Dicke model, including evidence of excited-state quantum phase transitions [7,35]. These advances also include microcanonical calculations and thermodynamic properties [36]. However, a complete picture of the quantum problem is still lacking. Alternatively, one can probe the properties of superradiant systems using inferences from stochastic processes. While limited in scope, the purpose of simplified models is to produce insights into specific aspects of the dynamics of superradiance in a more tractable manner [37]. Here, we address the evaluation of statistical moments and degeneracy in superradiance by considering a correspondence between a self-organizing process, which mimics the superradiance, and the well-known birthday problem. This connection allows one to study the degeneracy in both a small and large number of emitters and, thus, the aspects of the superradiant dynamics. The outline of the paper is as follows. In Section 2, we introduce a self-organizing process, in which a small initial fluctuation gives rise to the rapid growth of photon numbers in the cavity. The existing symmetries are identified, and analytic expressions for the ℓth statistical moments are derived. In Section 3, a correspondence between the self-organizing process and the birthday problem is unveiled allowing the exact evaluation of the ℓth statistical moment beyond the Poisson approximation (arbitrary time) and small N. We present a Monte Carlo example and compare its results to our exact calculations. Our closing remarks are listed in Section 4.

2. Superradiance

At its core, a superradiant pulse mimics a cascade of photon emission from a population composed of excited emitters. The complexity of the phenomenon arises from the coupling with a shared radiation field: the ensemble of N emitters can only create a single photon per unit of time, according to Equation (1). The constraint introduces temporal correlations among the emitters that dictate the collective decay. Here, we consider a self-organizing system formed by N emitters in a resonant cavity. Each emitter has two levels, and they are located close enough to each other to interfere through the common radiating field. At a given time instant t, there are $n(t)$ emitters in their respective ground states and $N - n(t)$ excited emitters. The self-organizing constraint is imposed by stating that no more than one emission occurs during a time interval δt, i.e., photon emission events are not independent. The emission process is composed by two stochastic events. A single emitter among N available is selected with uniform probability $p = 1/N$. If the selected emitter is found in the excited state, the subsequent emission occurs with conditional probability $p_e = 1$, otherwise $p_e = 0$. This stochastic process ensures that the total number of excited states can only be decremented by one for successive time steps. The assumption of uniform p implies that emitters are equally affected by the radiation field. It oversimplifies the spatial distribution of emitters in atomic clouds, excluding superradiant emission due to non-linear effects [37]. A naive analysis of the collective probability distribution, ignoring the superradiant constraint and thus correlations among emitters, leads to a Poissonian distribution. This is not the case, as we show in what follows.

Let $|n\rangle$ represent the configuration containing n emitters in the ground state in an ensemble with N emitters. The stochastic process compromises the transitions that occur along time on the vectors $|n\rangle$. The transitions are encoded by the transition matrix $\hat{T}$, which reads:

$$\hat{T} = \begin{pmatrix} 0 & 0 & 0 & & 0 & 0 \\ 1 & p & 0 & \cdots & 0 & 0 \\ 0 & 1-p & 2p & & 0 & 0 \\ & \vdots & & \ddots & \vdots & \\ 0 & 0 & 0 & \cdots & 1-p & 0 \\ 0 & 0 & 0 & & p & 1 \end{pmatrix}, \tag{2}$$

as illustrated in Figure 1. The operator $\hat{T}$ possesses some notable properties: $\text{Tr}(\hat{T}) = (1+p)/2$, and $\sum_{\mu=0}^{N} T_{\mu\nu} = 1$ ensures probability conservation. In addition, $\hat{T}$ is triangular with a clear pattern for each occupation level n. The analogy with angular momentum algebras leads to:

$$\hat{T} \equiv p(\hat{J}_z + j\hat{1}) + \hat{J}_+, \tag{3}$$

where $\hat{1}$ is the identity operator, $n = m_z + j$ ($j = 0, 1, \ldots, N/2$ and $m_z = -j, -j+1, \ldots, j$), $\hat{J}_+|n\rangle = (1-pn)|n+1\rangle$. Accordingly, $[\hat{J}^2, \hat{T}] = 0$, and j is a conserved quantity. The eigenvalue j is set by the initial conditions: for N emitters in the excited state, $j = N/2$.

Once the transition matrix is defined, one writes the master equation:

$$\partial_t |P(t)\rangle = -\hat{\mathcal{H}}|P(t)\rangle, \tag{4}$$

in which $\hat{\mathcal{H}} \equiv (\hat{1} - \hat{T})/\delta t$ is the generator of temporal translations, with a role similar to the Hamiltonian in quantum systems. The probability vector $|P(t)\rangle = P_0(t)|0\rangle + P_1(t)|1\rangle + \cdots + P_N(t)|N\rangle$ is a linear combination of occupation vectors, and the coefficients $P_n(t)$ describe the instantaneous probability to measure n emitters in the ground state. The coefficients $0 \leqslant P_n(t) \leqslant 1$ satisfy $\sum_{n=0}^{N} P_n(t) = 1$. The eigenvalues of $\hat{\mathcal{H}}$ are readily available, $E_n = (1 - pn)$, with $n = 0, 1, \ldots, N$, but the $\hat{\mathcal{H}}$ is not Hermitian. As a result, right and left eigenvectors, respectively $|\phi_k\rangle$ and $\langle\chi_k|$, are not related by Hermitian conjugation. The eigenvectors obey the identity decomposition, $\sum_{k=0}^{N} |\phi_k\rangle\langle\chi_k| = \hat{1}$, and are orthogonal to each other. Using the spectral decomposition, the solution of the stochastic problem reads $|P(t)\rangle = \sum_{n=0}^{N} c_n \mathrm{e}^{-E_n t}|\phi_n\rangle$, with $c_n = \langle\chi_n|P(0)\rangle$. The mode with vanishing eigenvalue E_N describes the asymptotic solution or stationary state, corresponding to all emitters in their respective ground states.

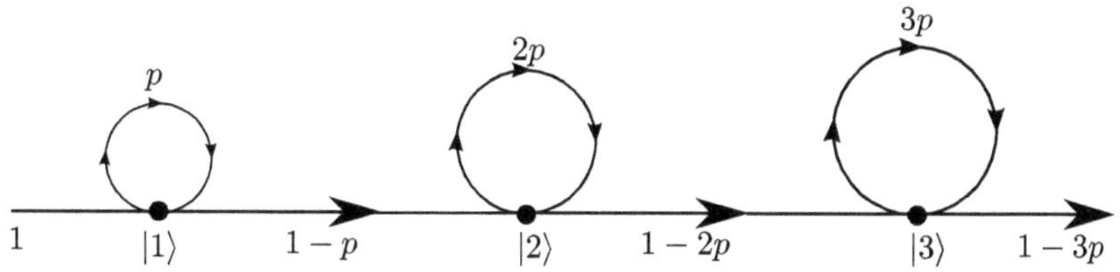

Figure 1. Superradiant stochastic process. The picture illustrates the transition $|n\rangle \to |n+1\rangle$ (horizontal arrows), with conditional probability $P(n+1, t+\delta t|n, t) = 1 - np$, and the transition $|n\rangle \to |n\rangle$ (loops), $P(n, t+\delta t|n, t) = np$. States with higher occupancy rates $n \gg 1$ have lower chances of radiating and spend more time in the same configuration. This process can be identified as the classical coupon collector's problem.

We propose an educated guess for left and right eigenvectors: there is a one-to-one correspondence between the coefficients of $\langle\chi_n|$ and the coefficients of the Pascal triangle of order N. Note that it is necessary to calculate only one set of eigenvectors, say $\{\langle\chi_n|\}$, because the remaining eigenvectors can be calculated via $\sum_{k=0}^{N} |\phi_k\rangle\langle\chi_k| = \hat{1}$. Surprisingly, the coefficients of $|\phi_n\rangle$ are also related to the Pascal triangle. For instance, the eigenvalues and respective right (left) eigenvectors are found in Table 1 (Table 2) with $N = 4$. The initial probability vector $|P(0)\rangle = |0\rangle$ acquires the following decomposition:

$$|P(0)\rangle = \sum_{n=0}^{N} \binom{N}{n} |\phi_n\rangle. \tag{5}$$

Using this result, we calculate the ℓth moment of the occupation number:

$$\langle n^{\ell}(t)\rangle = \mathrm{e}^{-t} \sum_{k=0}^{N} \sum_{n=k}^{N} \mathrm{e}^{(k/N)t} \binom{N}{k} \binom{N-k}{n-k} (-1)^{n+k} n^{\ell}\,. \tag{6}$$

so that the first statistical moment is $\langle n(t)\rangle = N(1 - \mathrm{e}^{-t/N})$. As expected, the system converges exponentially to a stationary state with a characteristic time scale $\tau = N$. Similarly, the second moment is evaluated,

$$\frac{\langle n^2(t)\rangle}{N^2} = 1 - \left(2 - \frac{1}{N}\right) \mathrm{e}^{-t/N} + \left(1 - \frac{1}{N}\right) \mathrm{e}^{-2t/N}, \tag{7}$$

while the standard deviation per emitter is:

$$\frac{\sigma}{N} = \frac{\mathrm{e}^{-t/2N}}{\sqrt{N}} + o(\mathrm{e}^{-3t/2N})\,. \tag{8}$$

Table 1. Right eigenvectors $|\phi_k\rangle = \sum_{n=0}^{N} d_n^{(k)} |n\rangle$ of $\hat{\mathcal{H}}$ with $N = 4$. The first row displays the eigenvalues E_k, while columns display the coefficients $d_n^{(k)}$ in the basis $|n\rangle$. Observe the combinatorial pattern in each column $d_n^{(k)} = (-1)^{k+n}(N-k)!/(n-k)!(N-n)!$ for $n \geq k$, 0 otherwise.

E	1	3/4	2/4	1/4	0
d_0	1				
d_1	−4	1			
d_2	6	−3	1		
d_3	−4	3	−2	1	
d_4	1	−1	1	−1	1

Table 2. Left eigenvectors $\langle \chi_k| = \sum_{n=0}^{N} \langle n| q_n^{(k)}$ of $\hat{\mathcal{H}}$ with $N = 4$. The first column displays the eigenvalues E_k, while rows display the coefficients $q_n^{(k)}$ in the basis $\langle n|$.

E	q_0	q_1	q_2	q_3	q_4
1	1				
3/4	4	1			
2/4	6	3	1		
1/4	4	3	2	1	
0	1	1	1	1	1

The crucial step that leads to Equation (6) is counting degenerate states derived from ordered permutations of N distinct objects. To do so, we consider the ordered time evolution (OTE) vector $|\psi(t_\kappa)\rangle \equiv |l_1 l_2 \cdots l_\kappa\rangle$, at discrete time t_κ. For the sake of convenience, $t_\kappa = \kappa\delta t$. Here, $l_j = 1, 2, \ldots, N$ identifies a single emitter selected at time t_j, with $j = 1, 2, \ldots, \kappa$. In this way, $|\psi(t_\kappa)\rangle$ is just a list containing all emitters selected along the time evolution. The number of distinct emitters selected during the interval $\kappa\delta t$ is extracted using the operator M, such that $M|\psi(t_\kappa)\rangle = m_\kappa|\psi(t_\kappa)\rangle$, with $m_\kappa = \sum_{n=1}^{N} \theta\left(\sum_{j=1}^{\kappa} \delta_{n,l_j}\right)$, where $\theta(x)$ is the step function $\theta(x > 0) = 1$, zero otherwise. Each eigenvalue m_κ is Ω_{κ,m_κ} degenerated. There are two discrete symmetries that leave M invariant: permutation of κ elements along time and permutation of N distinct objects. Both symmetries are expressed using the symmetric group $S_\kappa \otimes S_N$. Defining $\Omega_\kappa = \sum_{m=0}^{N} \Omega_{\kappa,m}$, the mean number of emitters in the ground state at time t_κ is:

$$\langle m_\kappa \rangle = \frac{1}{\Omega_\kappa} \sum_{m=0}^{N} m\, \Omega_{\kappa,m} \,. \tag{9}$$

We stress that the analytic calculations of the degeneracy $\Omega_{\kappa,m}$ become increasingly harder as either N or the time interval increases. Ultimately, the distribution $P_n(t)$ is the main goal, from which one can calculate or infer, for example, the instantaneous density of states or the instantaneous entropy $S(t) = -\sum_{n=0}^{N} P_n(t) \ln P_n(t)$. The values of $P_n(t)$ can be obtained either from statistical moments $\langle n^\ell(t)\rangle$ or from the instantaneous ratios $\Omega_{\kappa,n}/\Omega_\kappa$. However, it turns out that the calculation of $\Omega_{\kappa,m}$ can be simplified via a correspondence between the stochastic process and the classical birthday problem.

3. The Birthday Problem

Despite the hardships mentioned above, it is possible to craft a general solution using an elegant analogy with the classical process of drawing samples from an urn with N distinct emitters. One at a time, an emitter is drawn from the urn; its label is recorded in a list, and then, it is replaced in the urn. This process is repeated κ times. A sample is formed by the list of κ recorded labels. In each sample, let n be the number of distinct labels. Thus, $n = 1, \ldots, \min(\kappa, N)$. We call $\Omega'_{\kappa,n}$ the number of possible samples with n different labels.

This is precisely the formulation of the birthday problem, a set of κ elements and uniform randomly-chosen tags are assigned out of a set of N tags, with replacement, to each element. One concern is on the probability of having n distinct tags (out of N) assigned to the κ elements. There is a clear correspondence between the OTE configuration vectors in the limit $n \to m_\kappa$ and $\Omega'_{\kappa,n} \to \Omega_{\kappa,m_\kappa}$. The number of ways N distinct tags can be assigned to κ elements is N^κ. One can group the κ elements according to the n distinct tags. This counting is given by the Stirling number of the second kind, which is non-vanishing for $n \geq 1$ and written as:

$$\left\{ \begin{matrix} \kappa \\ n \end{matrix} \right\} = \frac{1}{n!} \sum_{j=0}^{n} (-1)^j \binom{n}{j} (n-j)^\kappa . \tag{10}$$

Furthermore, the number of possible distinct N tags in each group of size n is given by the falling factorial power, which we represent by the Pochhammer symbol $(N)_n = N(N-1)\ldots(N-n+1)$. The number of possible ways of grouping n distinct tags out of N in κ elements is precisely:

$$\Omega'_{\kappa,n} = (N)_n \left\{ \begin{matrix} \kappa \\ n \end{matrix} \right\} = \binom{N}{n} \sum_{j=0}^{n} (-1)^j \binom{n}{j} (n-j)^\kappa . \tag{11}$$

In addition, the total number of groupings reduces to a far simpler expression:

$$\Omega'_\kappa = \sum_{n=0}^{\min(\kappa,N)} \Omega'_{\kappa,n} = \sum_{n=0}^{\kappa} (N)_n \left\{ \begin{matrix} \kappa \\ n \end{matrix} \right\} = N^\kappa , \tag{12}$$

the upper limit in the summation can be taken to be κ since $\left\{ \begin{smallmatrix} \kappa \\ n \end{smallmatrix} \right\}$ vanishes for $n > \kappa$.

The interpretation using the finite permutation group is very appealing. For a fixed κ, the ℓth statistical moment of n is:

$$\langle n^\ell(t_\kappa) \rangle = \frac{\sum_{n=0}^{\min(\kappa,N)} \Omega'_{\kappa,n} n^\ell}{\sum_{n=0}^{\min(\kappa,N)} \Omega'_{\kappa,n}} = \frac{1}{N^\kappa} \sum_{n=0}^{\kappa} (N)_n \left\{ \begin{matrix} \kappa \\ n \end{matrix} \right\} n^\ell . \tag{13}$$

The degeneracy $\Omega'_{\kappa,n}$ is required to evaluate higher statistical moments in both microscopic and OTE approaches. However, the OTE results are valid for arbitrary time intervals between consecutive events, whereas the microscopic approach requires the temporal differential equation. Thus, OTE can properly describe the rapid fluctuations that would not be captured otherwise during the transient. For $\ell = 1$, one calculates that:

$$\langle n(t) \rangle = N \left[1 - \left(1 - \frac{1}{N} \right)^t \right] \approx N(1 - e^{-t/N}) , \tag{14}$$

and for $\ell = 2$,

$$\frac{\langle n^2(t) \rangle}{N^2} = \frac{\langle n(t) \rangle}{N} - \left(1 - \frac{1}{N} \right)^{t+1} \left[1 - \left(\frac{1 - 2/N}{1 - 1/N} \right) \right] . \tag{15}$$

Direct comparison between Equations (6) and (13) shows that both provide the same asymptotic $k = 1$ statistical moment. However, there is a striking difference: the OTE description in Equation (13) is based on the symmetries of finite groups, and it is also valid for arbitrary time intervals, whereas Equation (6) requires constant and small time intervals δt.

Figure 2 shows the time evolution of scaled standard deviation $\sigma(t)/\sqrt{N}$, for $N/32 = 100, 200, 500, 1000$, where all ensembles collapse under a single curve, dominated by the decay rate $\tau = 2N$ in discrete time. The standard deviation $\sigma(t)$ exhibits a maximum at $(t^*/N) \approx 1.3$, characterizing the rapid population change associated with the superradiant emission. However, these findings differ from the qualitative analysis of the Dicke model. In particular, $\langle n(t^*) \rangle \approx 0.73N > N/2$.

Considering $\delta t = 1/(Np_e)$, since at most a single emission can occur within the time interval δt, the average emitted power is $P_{\text{avg}} = \hbar\omega\langle n(t^*)\rangle/(t^*\delta t) \propto \hbar\omega N p_e$, which fails to recover the N^2 behavior expected for superradiance. This can be explained by noting that the transitions in the Dicke model increases approximately with nN until the maximum intensity is reached. Therefore, the simplification embedded in our model, that p_e remains constant along the process, slows downs the cascade of emissions and reduces the power of emitted radiation.

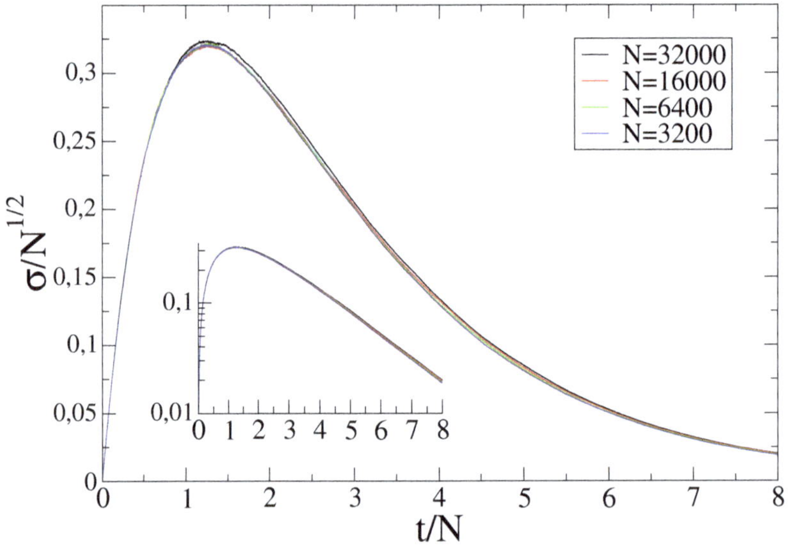

Figure 2. Time evolution of $\sigma(t)/\sqrt{N}$ using Monte Carlo simulations (50,000 runs). The natural time scale t/N was employed so that the scaled standard deviation $\sigma(t/N)/\sqrt{N}$ collapses under a single shared curve, for small and large N. The inset (log scale) displays the long time dominant behavior, which has a characteristic decay time $\tau = 2N$.

4. Conclusions

We have shown that the OTE approach can be used to compute high-order statistics in self-organizing systems. The increased dimensional space permits the identification of finite symmetries and novel recurrence relations, which are used to evaluate high-order statistics required in superradiant systems. Despite these advances, we stress that the model considered in this study oversimplifies the superradiant dynamics by assuming a fixed conditional probability of emission $p_e = 1$. An improved version of the model should contemplate p_e as a function of n, or at least consider an upper value $p_e^{\text{upper}} \propto N$, to address the N^2 behavior of the radiation intensity in superradiance. Finally, we also mention that the method can also be used in other stochastic problems such as the disease spreading by aerial vectors. In vector-borne diseases, pathogens are transmitted between humans or from animals to humans through bloodsucking insects. Examples of those diseases are: Dengue fever, yellow fever, malaria, and leishmaniasis, just to cite a few. Analogous to the process of drawing samples from an urn, each bite can be seen as drawn from an urn (population) with N distinct balls (hosts). Therefore, assuming that insect bites are randomly distributed among hosts, our result (Equation (6)) can be used to evaluate the statistical moments of this probabilistic system; where k is the number of bites, N the number of host individuals, and n is the number infected hosts.

Author Contributions: A.S.M., B.C., and G.M.N. designed the research; B.C. and G.M.N. performed the research and wrote the computational codes; A.S.M. verified the numerical results; B.C. and G.M.N. wrote the paper; A.S.M., B.C., and G.M.N. edited the paper; all authors reviewed the manuscript.

Funding: A.S.M. holds grants from CNPq 309851/2018-1. B.C. acknowledges Grant CAPES 88882.317469/2019-01. G.M.N. acknowledges Grant CAPES 88887.136416/2017-00.

Acknowledgments: We are grateful for César Augusto Terçariol and Tiago José Arruda for their helpful comments during the manuscript preparation.

Conflicts of Interest: The authors declare no conflict of interest.

Abbreviations

The following abbreviations are used in this manuscript:

DM	Dicke model
SOP	self-organizing phenomena
RWA	rotating wave approximation
OTE	ordered time evolution

References

1. Dicke, R.H. Coherence in Spontaneous Radiation Processes. *Phys. Rev.* **1954**, *93*, 99–110. [CrossRef]
2. Bohnet, J.G.; Chen, Z.; Weiner, J.M.; Meiser, D.; Holland, M.J.; Thompson, J.K. A steady-state superradiant laser with less than one intracavity photon. *Nature* **2012**, *484*, 78–81. [CrossRef] [PubMed]
3. Meiser, D.; Ye, J.; Carlson, D.R.; Holland, M.J. Prospects for a Millihertz-Linewidth Laser. *Phys. Rev. Lett.* **2009**, *102*, 163601. [CrossRef] [PubMed]
4. Meiser, D.; Holland, M.J. Intensity fluctuations in steady-state superradiance. *Phys. Rev. A* **2010**, *81*, 063827. [CrossRef]
5. Norcia, M.A.; Winchester, M.N.; Cline, J.R.K.; Thompson, J.K. Superradiance on the millihertz linewidth strontium clock transition. *Sci. Adv.* **2016**, *2*, [CrossRef] [PubMed]
6. Wang, D.W.; Liu, R.B.; Zhu, S.Y.; Scully, M.O. Superradiance Lattice. *Phys. Rev. Lett.* **2015**, *114*, 043602. [CrossRef] [PubMed]
7. Kloc, M.; Stránský, P.; Cejnar, P. Quantum quench dynamics in Dicke superradiance models. *Phys. Rev. A* **2018**, *98*, 013836. [CrossRef]
8. Scheibner, M.; Schmidt, T.; Worschech, L.; Forchel, A.; Bacher, G.; Passow, T.; Hommel, D. Superradiance of quantum dots. *Nat. Phys.* **2007**, *3*, 106–110. [CrossRef]
9. Mlynek, J.A.; Abdumalikov, A.A.; Eichler, C.; Wallraff, A. Observation of Dicke superradiance for two artificial atoms in a cavity with high decay rate. *Nat. Commun.* **2014**, *5*, 5186. [CrossRef]
10. Kozub, M.; Pawicki, L.; Machnikowski, P. Enhanced spontaneous optical emission from inhomogeneous ensembles of quantum dots is induced by short-range coupling. *Phys. Rev. B* **2012**, *86*, 121305. [CrossRef]
11. Abdussalam, W.; Machnikowski, P. Superradiance and enhanced luminescence from ensembles of a few self-assembled quantum dots. *Phys. Rev. B* **2014**, *90*, 125307. [CrossRef]
12. Yukalov, V.I.; Yukalova, E.P. Dynamics of quantum dot superradiance. *Phys. Rev. B* **2010**, *81*, 075308. [CrossRef]
13. Pustovit, V.N.; Shahbazyan, T.V. Cooperative emission of light by an ensemble of dipoles near a metal nanoparticle: The plasmonic Dicke effect. *Phys. Rev. Lett.* **2009**, *102*, 077401. [CrossRef]
14. Teperik, T.V.; Degiron, A. Superradiant Optical Emitters Coupled to an Array of Nanosize Metallic Antennas. *Phys. Rev. Lett.* **2012**, *108*, 147401. [CrossRef] [PubMed]
15. Yukalov, V.I. Origin of Pure Spin Superradiance. *Phys. Rev. Lett.* **1995**, *75*, 3000–3003. [CrossRef]
16. Yukalov, V.I.; Henner, V.K.; Kharebov, P.V.; Yukalova, E.P. Coherent spin radiation by magnetic nanomolecules and nanoclusters. *Laser Phys. Lett.* **2008**, *5*, 887. [CrossRef]
17. Yukalov, V.I.; Yukalova, E.P. Fast magnetization reversal of nanoclusters in resonator. *J. Appl. Phys.* **2012**, *111*. [CrossRef]
18. Yukalov, V.I.; Yukalova, E.P. Possibility of superradiance by magnetic nanoclusters. *Laser Phys. Lett.* **2011**, *8*, 804. [CrossRef]
19. Baumann, K.; Guerlin, C.; Brennecke, F.; Esslinger, T. Dicke quantum phase transition with a superfluid gas in an optical cavity. *Nature* **2010**, *464*, 1301–1306. [CrossRef]

20. Klinder, J.; Keßler, H.; Wolke, M.; Mathey, L.; Hemmerich, A. Dynamical phase transition in the open Dicke model. *Proc. Natl. Acad. Sci.* **2015**, *112*, 3290–3295. [CrossRef]
21. Baumann, K.; Mottl, R.; Brennecke, F.; Esslinger, T. Exploring Symmetry Breaking at the Dicke Quantum Phase Transition. *Phys. Rev. Lett.* **2011**, *107*, 140402. [CrossRef] [PubMed]
22. Hepp, K.; Lieb, E.H. On the superradiant phase transition for molecules in a quantized radiation field: the Dicke maser model. *Ann. Phys. (NY)* **1973**, *76*, 360–404. [CrossRef]
23. Rotondo, P.; Cosentino Lagomarsino, M.; Viola, G. Dicke Simulators with Emergent Collective Quantum Computational Abilities. *Phys. Rev. Lett.* **2015**, *114*, 143601. [CrossRef] [PubMed]
24. Nicolis, G.; Prigogine, I. Symmetry breaking and pattern selection in far-from-equilibrium systems. *Proc. Natl. Acad. Sci. USA* **1981**, *78*, 659–663. [CrossRef] [PubMed]
25. Witten, T.A. Insights from soft condensed matter. *Rev. Mod. Phys.* **1999**, *71*, S367–S373. [CrossRef]
26. Cross, M.C.; Hohenberg, P.C. Pattern formation outside of equilibrium. *Rev. Mod. Phys.* **1993**, *65*, 851–1112. [CrossRef]
27. Lima, G.F.; Martinez, A.S.; Kinouchi, O. Deterministic Walks in Random Media. *Phys. Rev. Lett.* **2001**, *87*, 010603. [CrossRef]
28. Cabella, B.C.T.; Martinez, A.S.; Ribeiro, F. Data collapse, scaling functions, and analytical solutions of generalized growth models. *Phys. Rev. E* **2011**, *83*, 061902. [CrossRef]
29. Berbert, J.M.; González, R.S.; Martinez, A.S. Ergodic crossover in partially self-avoiding stochastic walks. *Phys. Rev. E* **2013**, *88*, 032119. [CrossRef]
30. Tavis, M.; Cummings, F.W. Exact Solution for an N-Molecule—Radiation-Field Hamiltonian. *Phys. Rev.* **1968**, *170*, 379–384. [CrossRef]
31. Emary, C.; Brandes, T. Quantum Chaos Triggered by Precursors of a Quantum Phase Transition: The Dicke Model. *Phys. Rev. Lett.* **2003**, *90*, 044101. [CrossRef] [PubMed]
32. Brandes, T. Excited-state quantum phase transitions in Dicke superradiance models. *Phys. Rev. E* **2013**, *88*, 032133. [CrossRef]
33. Chen, Q.H.; Zhang, Y.Y.; Liu, T.; Wang, K.L. Numerically exact solution to the finite-size Dicke model. *Phys. Rev. A* **2008**, *78*, 051801. [CrossRef]
34. Liu, T.; Zhang, Y.Y.; Chen, Q.H.; Wang, K.L. Large-N scaling behavior of the ground-state energy, fidelity, and the order parameter in the Dicke model. *Phys. Rev. A* **2009**, *80*, 023810. [CrossRef]
35. Rodriguez, J.P.J.; Chilingaryan, S.A.; Rodríguez-Lara, B.M. Critical phenomena in an extended Dicke model. *Phys. Rev. A* **2018**, *98*, 043805. [CrossRef]
36. Bastarrachea-Magnani, M.A.; Lerma-Hernández, S.; Hirsch, J.G. Thermal and quantum phase transitions in atom-field systems: a microcanonical analysis. *J. Stat. Mech. Theory Exp.* **2016**, *2016*, 093105. [CrossRef]
37. Gross, M.; Haroche, S. Superradiance: An essay on the theory of collective spontaneous emission. *Phys. Rep.* **1982**, *93*, 301–396. [CrossRef]

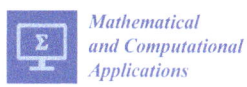

Mathematical and Computational Applications

Article

Inadequate Sampling Rates Can Undermine the Reliability of Ecological Interaction Estimation

Brenno Cabella [1], Fernando Meloni [2] and Alexandre S. Martinez [2,3,*]

[1] Instituto de Física Teórica, Universidade Estadual Paulista (UNESP), Rua Dr. Bento Teobaldo Ferraz 271, 01140-070 São Paulo, Brazil; cabellab@gmail.com
[2] Faculdade de Filosofia, Ciências e Letras de Ribeirão Preto (FFCLRP), Universidade de São Paulo (USP), Avenida Bandeirantes 3900, 14040-901 Ribeirão Preto, São Paulo, Brazil; fernandomeloni@usp.br
[3] Instituto Nacional de Ciência e Technologia de Sistemas Complexos (INCT-SC), Rua Dr. Xavier Sigaud 150, Urca, 22290-180 Rio de Janeiro, Brazil
* Correspondence: asmartinez@usp.br; Tel.: +55-16-3315-3720

Received: 28 March 2019; Accepted: 29 April 2019; Published: 30 April 2019

Abstract: Cycles in population dynamics are abundant in nature and are understood as emerging from the interaction among coupled species. When sampling is conducted at a slow rate compared to the population cycle period (aliasing effect), one is prone to misinterpretations. However, aliasing has been poorly addressed in coupled population dynamics. To illustrate the aliasing effect, the Lotka–Volterra model oscillatory regime is numerically sampled, creating prey–predator cycles. We show that inadequate sampling rates may produce inversions in the cause-effect relationship among other artifacts. More generally, slow acquisition rates may distort data interpretation and produce deceptive patterns and eventually leading to misinterpretations, as predators becoming preys. Experiments in coupled population dynamics should be designed that address the eventual aliasing effect.

Keywords: temporal aliasing effect; ecological methods; sampling rates; cyclic dynamics; predator–prey system; population biology

1. Introduction

Quantitative sampling provides the most important information source for ecological modeling. An important example, but not yet fully understood, is the periodic species abundance cycles in population dynamics. These cycles may appear in coupled systems, in which two or more elements (species or climate) interact in a cause-effect relationship. In Bulmer [1], lags between species cycles (phase shifts) were used to infer the relationship between different species in Canada.

Also, the historic data series observed by trappers working for Hudson's Bay Company, MacLulich [2] and Elton and Nicholson [3] found regular cycles in the population of Snowshoe Hares (*Lepus americanus*) and Canadian Lynx (*Lynx canadensis*). Their abundance have been matched, and showed an overlap with a small delay. The system was interpreted from the perspective of trophic interactions, as a regular predator-prey system, which was first labeled the Lotka–Volterra model (LVM) [4]. Some years later, the model became more robust, considering finite limits in the oscillatory predation rate [5]. Although predator–prey models are intuitively coherent and produce qualitative patterns found in nature, such models provide poor adjustment to field data, so their empiricism is still controversial [6].

There is a scientific consensus that better samples in field experiments lead to better interpretation of the real pattern. Effects caused by inappropriate sampling have already been addressed in the context of spatial influence on population dynamics or by the numerical insufficiency of samples [7,8]. However, period between samples are generally neglected and species interaction are especially prone to data misinterpretation when inappropriate sampling rates are used. The most common problem associated with slow sample rates is the aliasing effect. The usual concern of this artifact is the misidentification of a signal frequency [9–11]. However, aliasing may also occur in multivariate signals and its effect goes beyond frequency changes. In a bivariate coupled system the lag between prey and predators (phase shift) may also be compromised.

A thorough search in the scientific literature shows that the aliasing effect is poorly explored in Ecology, and its consideration may have deep implications. For instance, delays in coupled systems are ordinarily interpreted as competition effects in Ref. [12]. Also, Benicà and collaborators [13,14] have studied a long time series of plankton communities, applying regular samples to measure several species. The authors have found that the cause-effect relationship suggests a chaotic food web.

This manuscript is organized as follows: Firstly, we present the temporal aliasing effect and the Lotka–Volterra model. Next, we numerically solve the model and sample it with different rates. Finally, we present the results and artifacts due to poor sampling i.e., apparent inversion of cause-effect relationship, increased cycle period and synchronism.

2. Aliasing the Lotka–Volterra Model

Temporal aliasing effect occurs when the sampling rate is not fast enough compared to the system natural cycle period. For example, in movies, the spiked wheels on horse-drawn wagons sometimes appear to turn backwards, the "wagon-wheel effect", which is depicted in Figure 1. A wheel indeed turns clockwise, but due to the slow sampling by the camera (number of frames per second), a filmed wheel appears to turn counter-clockwise. This effect can be avoided considering the Nyquist–Shannon sampling theorem [15], which states that given a time series with minimum period τ, the equally spaced intervals between samples T_s must be smaller than half the minimum period, i.e., $T_s < \tau/2$.

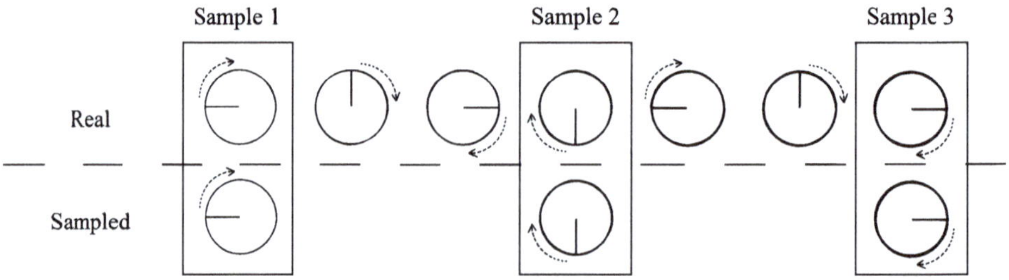

Figure 1. Aliasing wheel. Example of an aliasing effect in the clockwise rotation of a wheel. The visualized behavior on film is a counter-clockwise rotation, known as the wagon-wheel effect. The long time interval between samples explains this curiosity.

In Ecology, cycles are extensively found in systems in which species interact with each other and with the environment [6]. To illustrate how the aliasing effect may mislead the interpretation of population abundance cycles, consider a simple prey–predator interaction described by the LVM: $dx/dt = x(\alpha - \beta y)$ and $dy/dt = y(\delta x - \gamma)$, where $x(t)$ and $y(t)$ are the prey and predator population densities at time t, respectively, α is the prey growth rate in the absence of predators, γ is the predator death rate in the absence of prey, and β and δ are related to the interaction strength between both species. The LVM equations have two fixed points: the mutual extinction, $E_1(x^*, y^*) = (0, 0)$, and the neutral center, $E_2(x^*, y^*) = (\gamma/\delta, \alpha/\beta)$. Solutions around the singular point E_2 are cycles with period $\tau = 2\pi/\sqrt{\alpha\gamma}$. Although the LVM is not adequate to quantitatively describe real-world community dynamics, here it is suitable because of its cause-effect relation: the number of predators increases (decreases) after the prey abundance increases (decreases).

To demonstrate how sampling rates can change the patterns in predator–prey systems, we have numerically obtained the cyclic dynamic pattern using the LVM. The Lotka–Volterra differential equations have been implemented in MatLab® language, and their solutions have been obtained using the Dormand–Prince method [16]. Dormand–Prince is currently the default method in the **ode45** solver for MatLab®. The standard LV dynamics was obtained with the following parameters: $(x_0, y_0) = (1.01, 0.99)$, $\alpha = \beta = \delta = \gamma = 2\pi$ and $t \in [0, 1000]$ with resolution 10^{-3}. The model parameters have been set to produce a unitary oscillation period $\tau = 1$ near the equilibrium point E_2.

Figure 2a shows the prey (dashed line) and predator (full line) population cycles. Next, the prey (empty circle) and predators (filled circle) were sampled within fixed time intervals, T_s. We repeated the procedure, increasing T_s from $\tau/10$ until 1.1τ. For each sampling rate, we interpolated the points to build the respective time series to infer the original series. Based on the peaks of the time series, we inferred the oscillation period and the dephasing of predator and prey abundances. In all the cases, we considered all the individuals from both populations and sampling does not alter species interactions nor population densities. Moreover, any spacial effect is considered negligle. In this way, individuals are considered to be homogeneously spatially distributed (random mixing hypothesis), their number is large enough therefore we may neglect deviations around their mean densities, leading to the isolation of the sampling rate effect.

3. Sampling Effects

For $T_s < \tau/2$, the system real cycle period is correctly retrieved as expected by the Nyquist–Shannon theorem as displayed in Figure 2b, with $T_s = \tau/10$. As T_s increases, the signal becomes increasingly biased. Figure 2c,d depict different patterns even though the sample period is the same $T_s = 0.4$. In Figure 2e, $T_s = 0.48$, interleaved synchrony and anti-synchrony occurs for the same series. Even though the Nyquist–Shannon criteria is satisfied, i.e. the oscillation period of both species can be properly retrieved, the phase relation between them is disrupted. For exact $T_s = 0.5$, limiting value for the Nyquist–Shannon criteria, two possible behaviors emerge: In Figure 2f prey and predators abundances are anti-correlated; or perfectly correlated as shown Figure 2g. Similarly to the cases presented in Figure 2c,d, the observed series will depend on the initial sample, i.e., the phase of the cycle where the first sample is obtained. A further increase in T_s causes an inversion of the prey–predator cycle and also an enhancement of the population cycle period is observed. In Figure 2h, $T_s = 0.9$, an increasing (decreasing) in prey population is followed by a decrease (increase) in predators, which is the opposite expected from a prey–predator relationship.

The inverted cycle oscillations persist for even greater values of T_s as the oscillation period increases to $T_s \to \tau$. When $T_s = \tau$, there are no oscillations, as depicted in Figure 2i. For even greater values for T_s, the original dynamics is retrieved but with extended cycle period, shown in Figure 2j, $T_s = 1.1$. The patterns presented from Figure 2b–i repeat for $k\tau < T_s < (2k+1)\tau$ where $k = 0, 1, 2, \ldots$.

The effect of cycle inversion and frequency change is summarized in Figure 3. When prey and predator series are acquired with adequate sample rate, prey and predator abundances and lag between them are properly retrieved, see Figure 3a. Figure 3b presents the phase portrait in prey vs. predator plot where the cycle turns counterclockwise. However, when the sample rate is not adequate, an inversion of the cycle occurs, as shown in Figure 3c and seen in the phase portrait of Figure 3d. Figure 3e shows the changes in the observed frequency and prey–predator cycle direction as function of sample period T_s. Note that as T_s increases, the original direction of the cycle may be retrieved (counterclockwise) but the observed period (τ_o) will be always longer than the original.

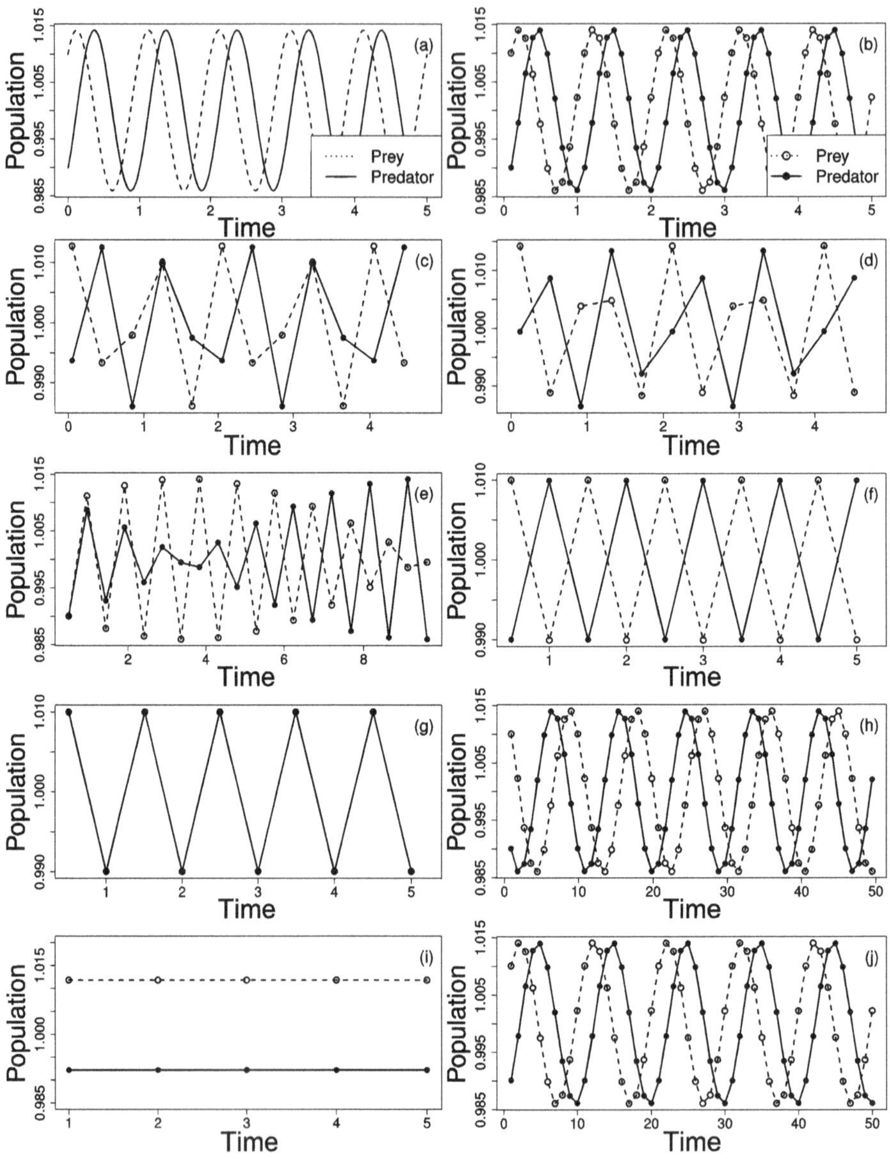

Figure 2. (**a**) Prey (dashed line) and predator (full line) population cycles obtained with the Lotka–Volterra model (LVM), with oscillation period $\tau = 1$. The LVM dynamics can generate different patterns due only to sampling rate effects. In the above panels, prey and predator abundances are represented by an empty and full circles, respectively. (**b**) $T_s = 0.1$, the time series correctly retrieve the LVM behavior. (**c**) $T_s = 0.4$, peaks seems to synchronize every two cycles. (**d**) $T_s = 0.4$, same sample period as in (**c**) but with different pattern. (**e**) $T_s = 0.48$, synchronous and anti-synchronous patterns are present in the same series. With $T_s = 0.5$ (Nyquist limit), two possible behaviors appears: (**f**) anti-synchronous cycles or (**g**) fully synchronized. (**h**) $T_s = 0.9$, an inversion and an extension of cycle period may be interpreted as preys eating predators. (**i**) $T_s = 1$ no oscillations are observed. (**j**) $T_s = 1.1$, the original dynamics is retrieved but with extended cycle period.

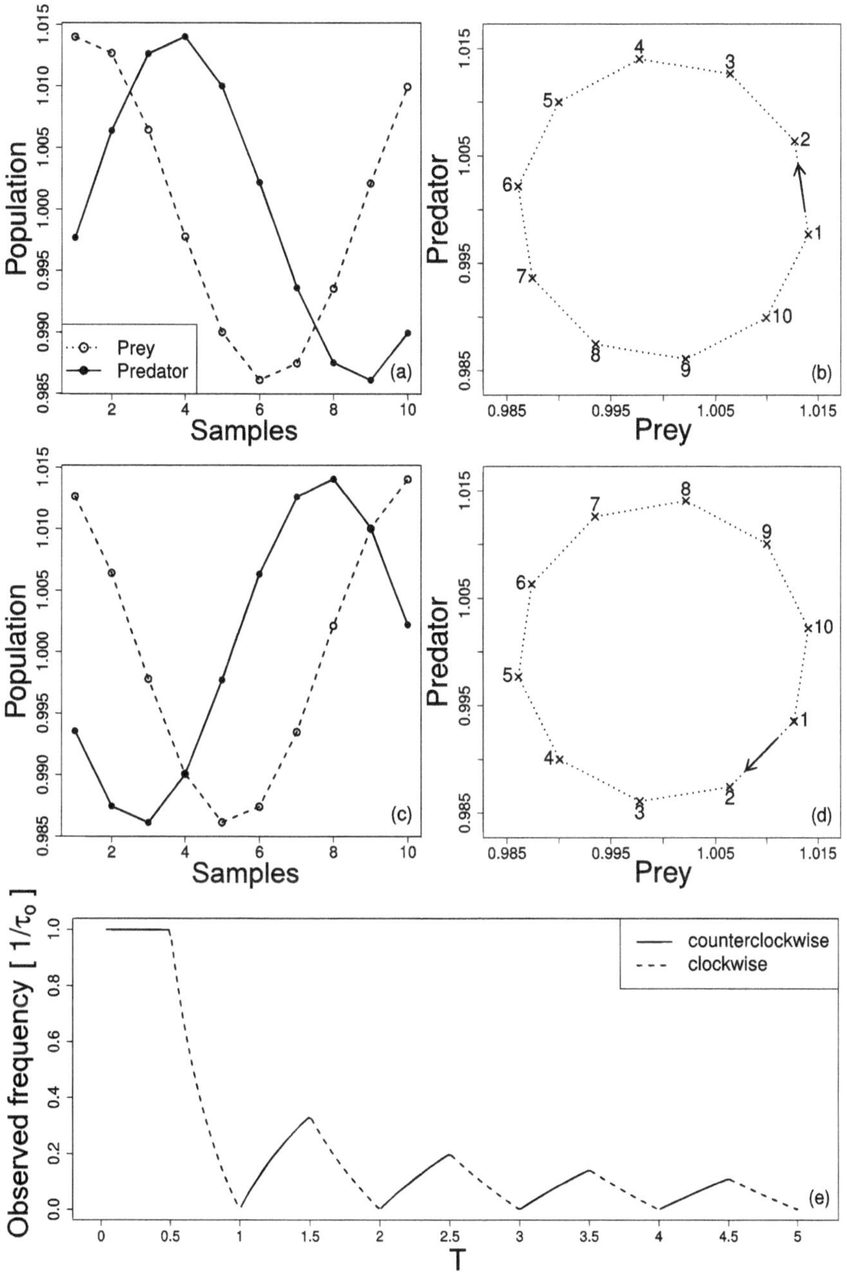

Figure 3. (**a**) Prey and predator series with adequate sample rate ($T_s = 0.1$) and corresponding phase portrait in (**b**). In a prey vs. predator plot, the cycle turns counterclockwise. (**c**) Prey and predator series with inadequate sample rate ($T_s = 0.9$) and corresponding phase portrait in (**d**). In this case the cycle turns clockwise. (**e**) changes in the observed frequency and prey–predator cycle direction as function of sample period T_s. Each sample in (**a**) and (**c**) is represented in (**b**) and (**d**) respectively.

A very simple and controlled oscillatory behavior, such the one LVM simulates, may produce different patterns in time series due only to inappropriate sampling rates, as shown in Figure 2b–g. Therefore, ecological interactions may be misinterpreted if data were collected with insufficiently sampling rates. In real world systems, this difficulty is amplified because the populations' periodicity is not necessarily constant and/or many species interactions may tangle the dynamics even more.

Aliasing should also be better evaluated in many other circumstances, such as the coupled aerosol-cloud-rain system, because the LVM is applied to modeling [17]. The influence of climate anomalies has been investigated as a driver of periods in population dynamics, as in the hare-lynx system [18,19]. Species abundance rates are the basis for evaluation of biological control success in crops, and in such cases, aliasing may have great financial consequences [20]. Sampling effects also have implications for biological conservation and species management, as in marine ecosystems, where population levels are used as a criterion to regulate fishing [21]. Further, some theoretical approaches about the trade-offs in Ecology and Evolution also concern predator-prey systems, trophic interactions or population cycles interactions [22–25]. Aliasing effect should also be considered on decision-making of public policies regarding national parks, fish stocks and hunting schedule, since the prediction of population levels often relies on sampled data.

4. Conclusion

We have stressed the importance of the aliasing effect in retrieving the behavior of ecological interactions. We have numerically demonstrated that slow sampling rates may lead to data misinterpretation. Aliasing is an often neglected effect that should be carefully considered when real data is used to model systems interaction. The aliasing hypothesis may provide new insights into old problems in Ecology and Biology. This result also highlights the importance of the field researchers, that can provide with realistic estimates for the population cycle periods, avoiding any circumstantial sampling with poor experimental designs.

Author Contributions: A.S.M., B.C. and F.M. designed the research; B.C. and F.M. performed the research and wrote computational codes; A.S.M. verified numerical results; B.C. and F.M. wrote the paper; A.S.M., B.C. and F.M. edited the paper. All authors reviewed the manuscript.

Funding: A.S.M. holds grants from CNPq 309851/2018-1, B.C. thanks Capes for support through a PNPD fellowship (88882.317469/2019-01) and F.M. acknowledges grant FAPESP 2013/06196-4.

Acknowledgments: We would like to thank the organizers of the Summer Course on Mathematical Methods in Population Biology, Roberto André Kraenkel and Paulo Inácio de Knegt López de Prado, where the hare-lynx paradox was first presented to us.

Conflicts of Interest: The authors declare no conflict of interest.

Abbreviations

The following abbreviation is used in this manuscript:

LVM	Lotka Voltera model

References

1. Bulmer, M.G. A Statistical Analysis of the 10-Year Cycle in Canada. *J. Anim. Ecol.* **1974**, *43*, 701–718. [CrossRef]
2. Barbosa, P.; Caldas, A.; Riechert, S.A. Species Abundance Distribution and Predator–Prey Interactions: Theoretical and Applied Consequences. In *Ecology of Predator–Prey Interactions*; Oxford University Press: Oxford, UK, 2005; pp. 344–368.
3. Elton, C.S.; Nicholson, M. The ten year cycle in numbers of lynx in Canada. *J. Anim. Ecol.* **1942**, *11*, 215–244. [CrossRef]
4. Odum, E.P. *Fundamentals of Ecology*; Saunders: Philadelphia, PA, USA, 1953.
5. Rosenzweig, M.L.; MacArthur, R.H. Graphical representation and stability conditions of predator–prey interactions. *Am. Nat.* **1963**, *97*, 209–223. [CrossRef]

6. Murray, J.D. *Mathematical Biology: I. An Introduction*; Springer: New York, NY, USA, 2012.
7. Kery, M.; Dorazio, R.M.; Soldaat, L.; van Strien, A.; Zuiderwijk, A.; Royle, J.A. Trend estimation in populations with imperfect detection. *J. Appl. Ecol.* **2009**, *46*, 1163–1172. [CrossRef]
8. Kishida, O.; Trussell, G.C.; Mougi, A.; Nishimura, K. Evolutionary ecology of inducible morphological plasticity in predator–prey interaction: Toward the practical links with population ecology. *Popul. Ecol.* **2010**, *52*, 37–46. [CrossRef]
9. Oppenheim, A.V.; Schafer, R.W. Sampling of continuous-time signals. In *Discrete-Time Signal Processing*; Prentice Hall: Upper Saddle River, NJ, USA, 1999; pp. 140–150.
10. Green, D.G. Time Series and Postglacial Forest Ecology. *Quat. Res.* **1981**, *15*, 265–277. [CrossRef]
11. Ford, D.E.; Thornton, K.W. Time and length scales for the one-dimensional assumption and its relation to ecological models. *Water Resour. Res.* **1979**, *15*, 113–120. [CrossRef]
12. Vandermeer, J. Coupled oscillations in food webs: Balancing competition and mutualism in simple ecological models. *Am. Nat.* **2004**, *163*, 857–867. [CrossRef] [PubMed]
13. Benincà, E.; Huisman, J.; Heerkloss, R.; Jöhnk, K.D.; Branco, P.; van Nes, E.H.; Scheffer, M.; Ellner, S.P. Chaos in a long-term experiment with a plankton community. *Nature* **2008**, *451*, 822–825. [CrossRef] [PubMed]
14. Benincà, E.; Jöhnk, K.D.; Heerkloss, R.; Huisman, J. Coupled predator–prey oscillations in a chaotic food web. *Ecol. Lett.* **2009**, *12*, 1367–1378. [CrossRef] [PubMed]
15. Nyquist, H. Certain Topics in Telegraph Transmission Theory. *Trans. Am. Inst. Electr. Eng.* **1928**, *47*, 617–644. [CrossRef]
16. Dormand, J.R.; Prince, P.J. A family of embedded Runge–Kutta formulae. *J. Comput. Appl. Math.* **1980**, *6*, 19–26. [CrossRef]
17. Koren, I.; Feingold, G. Aerosol-cloud-precipitation system as a predator–prey problem. *Proc. Natl. Acad. Sci. USA* **2011**, *108*, 12227–12232. [CrossRef] [PubMed]
18. Yan, C.; Stenseth, N.C.; Krebs, C.J.; Zhang, Z. Linking climate change to population cycles of hares and lynx. *Glob. Chang. Biol.* **2013**, *19*, 3263–3271. [CrossRef]
19. Stenseth, N.C. Canadian hare–lynx dynamics and climate variation: Need for further interdisciplinary work on the interface between ecology and climate. *Clim. Res.* **2007**, *34*, 91–92. [CrossRef]
20. Snyder, W.E.; Chang, G.C.; Prasad, R.P. Conservation Biological Control: Biodiveristy Influences the Effectiveness of Predators. In *Ecology of Predator–Prey Interactions*; Oxford University Press: Oxford, UK, 2005; p. 324.
21. Hunsicker, M.E.; Ciannelli, L.; Bailey, K.M.; Buckel, J.A.; Wilson White, J.; Link, J.S.; Essington, T.E.; Gaichas, S.; Anderson, T.W.; Brodeur, R.D.; et al. Functional responses and scaling in predator–prey interactions of marine fishes: Contemporary issues and emerging concepts. *Ecol. Lett.* **2011**, *14*, 1288–1299. [CrossRef] [PubMed]
22. Weitz, J.S.; Levin, S.A. Size and scaling of predator–prey dynamics. *Ecol. Lett.* **2006**, *9*, 548–557. [CrossRef]
23. Cortez, M.H. Comparing the qualitatively different effects rapidly evolving and rapidly induced defences on predator–prey Interactions. *Ecol. Lett.* **2011**, *14*, 202–209. [CrossRef] [PubMed]
24. Kalinkat, G.; Schneider, F.D.; Digel, C.; Guill, C.; Rall, C.B.; Brose, U. Body masses, functional responses and predator–prey stability. *Ecol. Lett.* **2013**, *16*, 1126–1134. [CrossRef] [PubMed]
25. Schneider, F.D.; Scheu, S.; Brose, U. Body mass constraints on feeding rates determine the consequences of predator loss. *Ecol. Lett.* **2012**, *15*, 436–443. [CrossRef] [PubMed]

Mathematical and Computational Applications

MDPI

Article

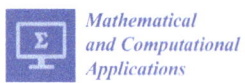

Finite Symmetries in Agent-Based Epidemic Models

Gilberto M. Nakamura [1,2], Ana Carolina P. Monteiro [1], George C. Cardoso [1] and Alexandre S. Martinez [1,2,*]

1 Faculdade de Filosofia, Ciências e Letras de Ribeirão Preto (FFCLRP), Universidade de São Paulo (USP), Avenida Bandeirantes 3900, 14040-901 Ribeirão Preto, São Paulo, Brazil; gmnakamura@usp.br (G.M.N.); ana.carolina.monteiro@usp.br (A.C.P.M.); gcc@usp.br (G.C.C.)

2 Instituto Nacional de Ciência e Technologia de Sistemas Complexos (INCT-SC), Rua Dr. Xavier Sigaud 150, Urca, 22290-180 Rio de Janeiro, Brazil

* Correspondence: asmartinez@usp.br; Tel.: +55-16-3315-3720

Received: 1 March 2019; Accepted: 21 April 2019; Published: 23 April 2019

Abstract: Predictive analysis of epidemics often depends on the initial conditions of the outbreak, the structure of the afflicted population, and population size. However, disease outbreaks are subjected to fluctuations that may shape the spreading process. Agent-based epidemic models mitigate the issue by using a transition matrix which replicates stochastic effects observed in real epidemics. They have met considerable numerical success to simulate small scale epidemics. The problem grows exponentially with population size, reducing the usability of agent-based models for large scale epidemics. Here, we present an algorithm that explores permutation symmetries to enhance the computational performance of agent-based epidemic models. Our findings bound the stochastic process to a single eigenvalue sector, scaling down the dimension of the transition matrix to $o(N^2)$.

Keywords: Markov processes; computational methods; epidemic models; complex systems; nonlinear dynamics

1. Introduction

In recent years, the emergence of Zika and Ebola viruses have attracted much attention from the scientific community after reports of their aggressive effects, respectively, microcephaly in newborns [1] and high mortality rate [2–4]. Despite their intrinsic differences concerning transmission mechanisms and pathogen-host interaction, both viruses spread in a population starting from a few infected individuals, based on their geographic location and network of contacts. Contact tracing and proper clinical care planning are critical parts of the WHO strategic plan [5] to mitigate ongoing transmissions and incidence cases, requiring the correct spatiotemporal dissemination of the disease. This assertion has renewed the interest in agent-based epidemic models (ABEM).

ABEM are mathematical models that describe the evolution of infectious diseases among a finite number N of agents over time (see Reference [6] for an extensive review). For that purpose, agents are labeled using integer numbers $k = 0, 1, \ldots, N-1$, whereas contacts between agents are mapped via an adjacency matrix A. The matrix elements are $A_{ij} = 1$ if the j-th agent connects to the i-th agent and otherwise vanishes. Accordingly, the set formed by agents and their interconnection is expressed as a graph, as depicted in Figure 1. In this way, heterogeneity arises naturally since the individuality of agents is taken into account, distinguishing ABEM from compartmental epidemic models [7–9].

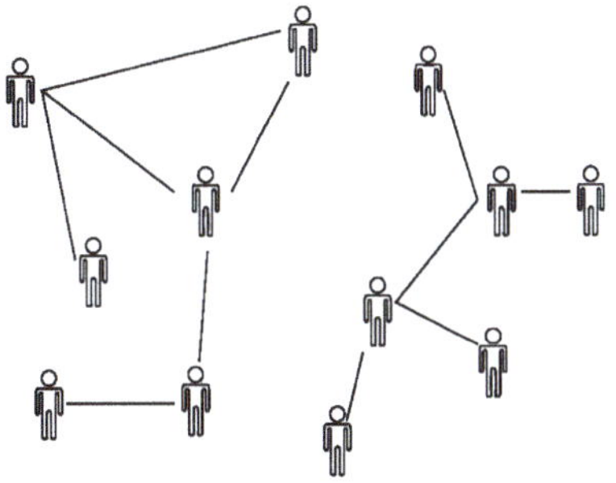

Figure 1. Agent network. Agents (vertices) and their interconnections (edges) are expressed as a graph. The graph representation introduces heterogeneity among the agents, which must be accounted for during disease spreading.

The susceptible-infected-susceptible model (SIS) is the simplest ABEM. It considers only two health states for agents, infected $|1\rangle$ or susceptible $|0\rangle$, and the occurrence of the following events during a time interval δt [10,11]. An infected agent may undergo a recovery event and return to susceptible state with probability γ; an infected agent may infect a susceptible agent with transmission probability β if and only if both agents are connected; or remains unchanged, as Figure 2 illustrates. Therefore, the SIS ABEM is inherently a Markov process in discrete time. The time interval δt is often chosen so that sequential recovery-recovery or transmission-recovery events are unlikely within the available time window. This is the so-called Poissonian hypothesis [12–15].

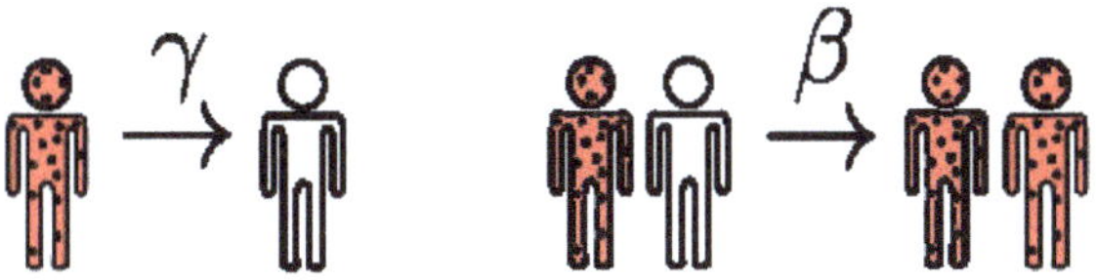

Figure 2. Susceptible-infected-susceptible model (SIS) transition events. Infected agents (red dotted) undergo recovery events with probability γ and change to susceptible (empty) health status. Infected agents may also infect additional susceptible agents with probability β, as long they are connected.

Following Reference [16], any configuration of N agents is obtained by direct composition of individual agent states. Let μ be an integer that labels the μ-th configuration so that:

$$|\mu\rangle \equiv |n_{N-1} \cdots n_1\, n_0\rangle \,, \tag{1}$$

with $n_k = 0, 1$ and $\mu = n_{N-1}2^{N-1} + \cdots + n_0 2^0$. A simple example for $N = 4$ is $|8\rangle \equiv |1000\rangle$, which represents the configuration where only the agent $k = 3$ is infected. From this scheme, it is already clear that there exist 2^N configurations in total since there are two available states for each agent. In what follows, we employ the notation: Latin indices enumerate agents $0, 1, \ldots, N-1$, while Greek indices enumerate configuration states $0, 1, \ldots, 2^N - 1$.

Let $|\pi(t)\rangle$ be the probability vector and $\pi_\mu(t) = \langle\mu|\pi(t)\rangle$ the probability of observing the configuration $|\mu\rangle$ at time t [17,18]. The master equation for the general Markov process reads:

$$\frac{d}{dt}\pi_\mu(t) = -\sum_\nu H_{\mu\nu}\pi_\nu(t), \tag{2}$$

$\hat{H} = (\mathbb{1} - \hat{T})/\delta t$ is the step operator, whereas $\hat{T}$ stands for the transition matrix [19]. The transition matrix $\hat{T}$ encodes all transitions available between any two configuration vectors. Its matrix elements $T_{\mu\nu}$ are constructed from much simpler rules. These rules are model dependent and fully characterize the stochastic model, as we shall see in details later. If a representation of $\hat{T}$ is known, the solution of the master equation provides the instantaneous values of the probabilities $\pi_\mu(t)$, i.e., the entire probability distribution function. Therefore, it becomes possible to calculate any relevant statistics of the problem at any instant of time, including those that may not be easily accessible or accurate by other numerical methods, such as the instantaneous Shannon entropy or the characteristic function.

Luckily, for time independent $\hat{T}$, the solution is well known:

$$|\pi(t)\rangle = \mathrm{e}^{-\hat{H}t}|\pi(0)\rangle \ . \tag{3}$$

Despite the existence of this exact solution, the applicability of Equation (3) at this stage is limited to small population sizes $N \sim O(20)$. The reason is the exponential growth of the underlying vector space with N. Here, we present algorithms to generate the operators $\hat{T}$ and $\hat{H}$ using finite symmetries or, equivalently, permutation symmetries via Cayley's theorem [20]. These algorithms are usually applied to condensate matter physics [21,22], but they may also be employed in epidemiology studies, due to recent developments in the disease spreading dynamics [16]. For pedagogical reasons, we first show how to build the complete 2^N vector space and the corresponding transition matrix. Next, we explore cyclic permutations to construct the cyclic vector space, in which $\hat{T}$ is broken down into N smaller blocks. Lastly, we consider the most symmetric cases, which reduce the problem to O(N). These instances correspond to the mean field or averaged networks. The iteration of sparse $\hat{T}$ over $|\pi(t)\rangle$ produces the desired disease evolution among agents. Relevant steps are shown in Algorithm A1..

2. Transition Matrix

The transition matrix $\hat{T}$ for an SIS model considering N two-state agents is [16]:

$$\hat{T} = \mathbb{1} - \beta \sum_{kj} \left[A_{jk}(1 - \hat{n}_j - \hat{\sigma}_j^+) + \Gamma \delta_{kj}(1 - \hat{\sigma}_j^-) \right] \hat{n}_k \ , \tag{4}$$

where $\Gamma = \gamma/\beta$, δ_{kl} is the Kronecker delta, $\hat{n}_k|n_k\rangle = n_k|n_k\rangle$, is the local number operator ($n_k = 0, 1$), and $\hat{\sigma}_k^+|n_k\rangle = \delta_{n_k,0}|1_k\rangle$, $\hat{\sigma}_k^-|n_k\rangle = \delta_{n_k,1}|0_k\rangle$ are the Pauli raising and lowering local operators, respectively. Local algebraic relationships are $[\hat{n}_k, \hat{\sigma}_{k'}^{\pm}] = \pm\delta_{k'k}$ and $[\hat{\sigma}_k^+, \hat{\sigma}_{k'}^-] = \delta_{k'k}(2\hat{n}_k - \mathbb{1})$. Inspection of Equation (4) readily shows $\hat{T}$ is not Hermitian. This means left- and right-eigenvectors are not related by Hermitian conjugation. In this scenario, the correct time evolution of $\pi_\mu(t)$ using Equation (3) requires the complete eigendecomposition, i.e., 2^N eigenvalues accompanied by 2^N right-eigenvectors and 2^N left-eigenvectors. This is often the main criticism against ABEM [12].

However, the scenario described above is not entirely correct. The rationale behind it assumes all eigenstates are equally relevant, which is incorrect whenever A exhibits invariance upon the action of a particular group (sets of transformations). Symmetries allow the matrix representation of $\hat{T}$ to be in block diagonal form, as depicted in Figure 3. Eigenvectors related to each block share the same eigenvalue (degeneracy), as usual in quantum mechanics [23]. Therefore, the trick to simplify problems involving the transition matrix lies in the selection of the appropriate basis in respect to a given symmetry, creating matrix representations with smaller blocks. The computational performance using this methodology surpasses that of working with the full matrix because each block can be treated separately, reducing memory storage and access. In particular, if only a few blocks are relevant to the analysis, the remaining blocks can be neglected. This property often produces massive reductions in the typical dimensions of the problem, enhancing computation times.

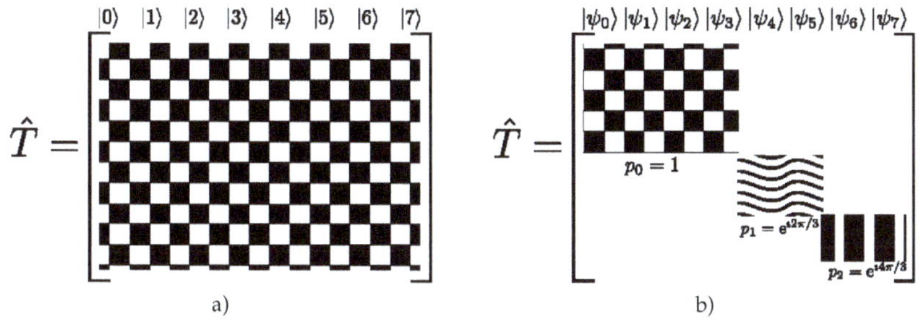

Figure 3. Reduction of the transition matrix to block diagonal form. (**a**) In the configurational vector space, $\{|\mu\rangle\}$, the matrix representation of $\hat{T}$ lacks an explicit mathematical pattern. (**b**) The emergence of organizational patterns are observed whenever symmetries of $\hat{T}$ are correctly addressed by employing the eigenvectors $\{\psi\}$ and eigenvalues $\{\lambda\}$ corresponding to the symmetry group considered. Under the invariant basis $\{\psi\}$, the matrix representation of $\hat{T}$ is brought to a block diagonal form, with blocks labeled by eigenvalues $\{\lambda\}$.

In the SIS model, recovery events result from actions of one-body operators, $\hat{\sigma}_k^- \hat{n}_k \equiv \hat{\sigma}_k^-$, on configuration vectors. Infection events are two-body operators: one infected agent may transmit the communicable disease to a susceptible agent after interaction between them, in the time interval δt. Interestingly, the resulting interaction also depends on symmetries available to the adjacency matrix A. The symmetries available to A may be further explored to assemble the initial vector space, reducing $\hat{T}$ to its block diagonal form.

Group operations over A are always finite transformations. One may explore the isomorphism between finite groups and the permutation group via Cayley theorem [20] to build permutation invariant subspaces. To that end, one must select the finite group and the corresponding symmetry. For graphs, the circular representation provides a convenient context to explore the existing symmetries, as Figure 4a depicts. From Figure 4b, connections among agents remain unchanged after cyclic permutation of agents, hence, A exhibits invariance under cyclic permutations. Cyclic permutations form a subset of permutation group and often represent geometric transformations, such as rotations and translations.

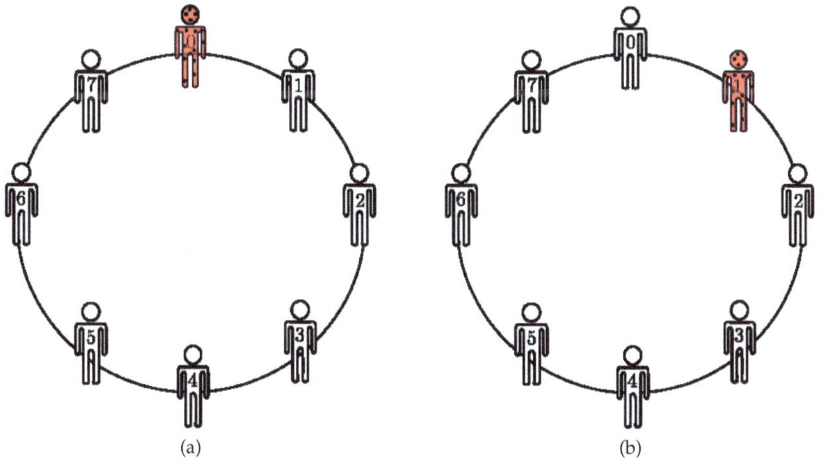

Figure 4. Regular graph in circular representation for $N = 8$ and single infected agent (red dotted). (**a**) The infected agent lies at node $k = 0$. (**b**) Graph obtained from cyclic permutation of nodes $k \to k + 1$ and $N - 1 \to 0$. Connections remain unchanged.

Vectors with N agents and invariant by cyclic permutations are built as follows. Consider the *representative* vector:

$$|\mu_p\rangle \equiv \frac{1}{\mathcal{N}_\mu} \sum_{k=0}^{N-1} \left(e^{2i\pi p/N}\hat{P}\right)^k |\mu\rangle \,, \tag{5}$$

where $\mathcal{N}_\mu$ is the normalization and $\hat{P}$ is the single step cyclic permutation with $p = 0, 1, \ldots, N-1$. The eigenvalues $e^{-2i\pi p/N}$ are derived from $\hat{P}^N = \mathbb{1}$. Eigenvalues can be associated with invariant subspaces, or sectors, spanned by their corresponding eigenvectors. For the sake of convenience, the integer p labels the eigenvalue sector. The representative vector $|\mu_p\rangle$ describes the linear combination of N-agent configurations related to $|\mu\rangle$ by cyclic permutations. For instance, $|3_0\rangle = (|011\rangle + |110\rangle + |101\rangle)/\sqrt{3}$ corresponds to the representative vector for $\mu = 3$, with $N = 3$ in the $p = 0$ sector. By construction, the vectors $|\mu_p\rangle$ satisfy the eigenvalue equation $\hat{P}|\mu_p\rangle = e^{-2i\pi p/N}|\mu_p\rangle$. They are also useful to identify symmetries, as they never change link distributions, only node labels. If $\hat{T}$ is symmetric under cyclic permutations, $\hat{T}$ and $\hat{P}$ commute with each other $[\hat{T}, \hat{P}] = 0$, meaning they share a common set of eigenvectors. Thus, $\hat{T}$ can be written using $|\mu_p\rangle$ and, more importantly, transitions between eigenvectors with distinct eigenvalues are prohibited. This feature leads to a block diagonal form to the matrix representation of $\hat{T}$.

3. Cyclic Vector Space

The complete picture of infection dynamics generated by the SIS model requires the utilization of 2^N configuration vectors. For completeness sake, we discuss the algorithm to obtain the vector space using both string and numeric representations. Matrix elements of $\hat{T}$ in Equation (4) are calculated from an adjacency matrix and user input dictionary (lookup table) based on off-diagonal transition rules.

According to Equation (1), the configuration vector $|\mu\rangle$ is obtained from the binary representations of the labels μ, as exemplified in Figure 5. There are two common equivalent routes to implement the configuration in computer codes. The first method employs string objects whereas the second method makes use of discrete mathematics. The second approach tends to be more efficient for two-state problems as optimized and native libraries for binary operations are widely available. For pedagogical purposes and generalization for more than two-states, we avoid exclusive binary operations in favor of usual discrete integer division and modulo operations.

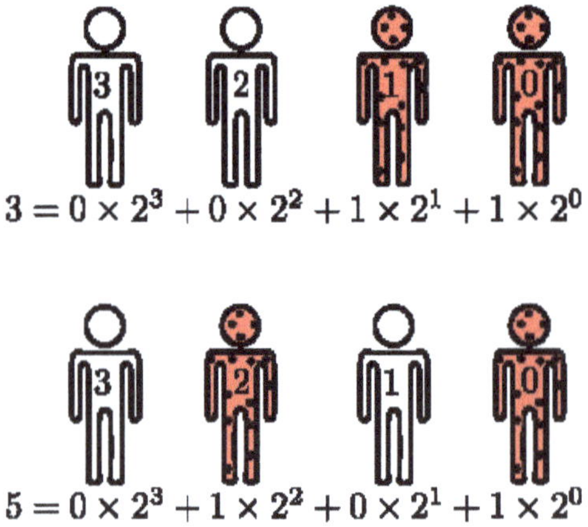

Figure 5. Agent configurations using binary representation for $\mu = 3$ and 5 with $N = 4$. For $|\mu = 3\rangle = |0011\rangle$, whereas $|\mu = 5\rangle = |0101\rangle$. In both configurations, two agents are infected (red dotted).

In Python programming language, classes provide a convenient mechanism to enable both formalisms for each instanced object (vector). Here, the custom class SymConf is used to encapsulate two instance variables: *label* stores the string representation of N agents, while *label_int* stores the corresponding integer number. In addition, the custom class also encapsulates three global class variables, *base*, *dimension*, and *basemax*, whose default values are 2, N, and 2^N. Base corresponds to the number of available states per agent. The class main method generates the eigenvectors $|\mu_p\rangle$ with eigenvalue $\exp(-2i\pi p/N)$, relative to the cyclic permutation operator $\hat{P}$ using Equation (5).

In what follows, we address four relevant points regarding the permutation eigenvectors $|\mu_p\rangle$, namely, the criteria used to label eigenvectors; normalization; number of infected agents; and the permutation operation.

Labels. Equation (5) claims permutation eigenvectors are a linear combination of all configuration vectors related by cyclic permutations. Here, we set the convention to adopt the smallest value μ present in the linear combination to label the representative vector. As examples, consider the following representatives of $\mu = 1$, $N = 4$, and $p = 0, 1, 2, 3$:

$$|1_0\rangle = \frac{|0001 = 1\rangle + |0010 = 2\rangle + |0100 = 4\rangle + |1000 = 8\rangle}{\sqrt{4}}. \tag{6a}$$

$$|1_1\rangle = \frac{|1\rangle + i|2\rangle - |4\rangle - i|8\rangle}{\sqrt{4}}. \tag{6b}$$

$$|1_2\rangle = \frac{|1\rangle - |2\rangle + |4\rangle - |8\rangle}{\sqrt{4}}. \tag{6c}$$

$$|1_3\rangle = \frac{|1\rangle - i|2\rangle - |4\rangle + i|8\rangle}{\sqrt{4}}. \tag{6d}$$

The order convention is necessary to calculate the relative phase between configurations related by permutations in non-trivial linear combinations. For instance, consider the vector $|\phi\rangle = \hat{P}|\mu_p\rangle = (1/\mathcal{N}_\mu)\sum_k (e^{2i\pi p/N}\hat{P})^k \hat{P}|\mu\rangle$. Since $|\mu_p\rangle$ and $|\phi\rangle$ are related by a single cyclic permutation, they differ by a phase factor: $|\phi\rangle = e^{-2i\pi p/N}|\mu_p\rangle$. Note that the linear combination $\hat{P}|\mu_p\rangle + |\mu_p\rangle = (1 + e^{-2i\pi p/N})|\mu_p\rangle$ vanishes for $p = N/2$. Despite the simplicity of the previous example, it already illustrates the relevance of phase difference among cyclic vectors.

Normalization. According to Equation (5), the squared norm of representative vectors is:

$$\langle \mu_p | \mu_p \rangle = \frac{1}{\mathcal{N}_\mu} \sum_{k=0}^{N-1} e^{-2i\pi pk/N} \langle \mu | \hat{P}^{-k} | \mu_p \rangle = \frac{N}{\mathcal{N}_\mu} \langle \mu | \mu_p \rangle \,. \tag{7}$$

The evaluation of the scalar product $\langle \mu | \mu_p \rangle$ follows directly from Equation (5). One notices the configuration $|\mu\rangle$ may appear only once for several linear combinations $|\mu_p\rangle$, so that $\langle \mu | \mu_p \rangle = 1/\mathcal{N}_\mu$. For instance, this is the case of $\langle 1 | 1_p \rangle$. However, a given configuration $|\mu\rangle$ may contribute more than once if there exists an integer $1 \leq r \leq N$ such that $\hat{P}^r|\mu\rangle = |\mu\rangle$, i.e., after r cyclic permutations the configuration repeats itself. Since $\hat{P}^N = \mathbb{1}$, it follows N/r is the number of times the configuration $|\mu\rangle$ appears in $|\mu_p\rangle$. Each contribution adds $e^{2i\pi pm/N}/\mathcal{N}_\mu$ $(m = 0, 1, \ldots, N/r - 1)$ in Equation (7). This result is conveniently summarized using the repetition number:

$$R_{\mu,p} = \sum_{m=0}^{N/r-1}{}' \,(e^{2i\pi pr/N})^m, \tag{8}$$

where the primed sum indicates N/r in the upper limit is an integer number. Therefore, $\langle \mu | \mu_p \rangle = R_{\mu,p}/\mathcal{N}_\mu$ and one obtains $\mathcal{N}_\mu = \sqrt{NR_{\mu,p}}$ from Equation (7).

We now show two examples to consolidate the discussion around $R_{\mu,p}$ and $\mathcal{N}_\mu$, for $N = 4$ and two infected agents. The configuration state $|3\rangle = |0011\rangle$ requires N cyclic permutations to repeat itself, so that $R_{3,p} = 1$ for any p and the corresponding normalization for $|3_p\rangle$ is simply $\mathcal{N}_3 = \sqrt{N}$, as expected. The first non-trivial case arises for $|5_p\rangle$ because the base configuration $|5\rangle = |0101\rangle$ satisfies $\hat{P}^2|5\rangle = |5\rangle$. According to Equation (8), $R_{5,p} = 1 + e^{4i\pi p/N}$ and assume only values: $R_{5,0} = R_{5,2} = 2$ and $R_{5,1} = R_{5,3} = 0$. Thus, depending on p, certain linear combinations are *forbidden* because they produce vectors with null norm, ensuring the correct dimension of vector space. The remaining non-null states for $N = 4$ are shown in Table 1 for further reference.

Table 1. Cyclic permutation eigenvectors with $N = 4$ agents. The first column shows the number of infected agents in the eigenvector. Each remaining column corresponds to a permutation sector p, and each row the corresponding state $|\mu_p\rangle$. The cross symbol indicates null-normed vector and the dimension of the vector space is $d = 2^4$.

n	$p=0$	$p=1$	$p=2$	$p=3$
0	$\|0_0\rangle$	$\times$	$\times$	$\times$
1	$\|1_0\rangle$	$\|1_1\rangle$	$\|1_2\rangle$	$\|1_3\rangle$
2	$\|3_0\rangle$	$\|3_1\rangle$	$\|3_2\rangle$	$\|3_3\rangle$
2	$\|5_0\rangle$	$\times$	$\|5_2\rangle$	$\times$
3	$\|7_0\rangle$	$\|7_1\rangle$	$\|7_2\rangle$	$\|7_3\rangle$
4	$\|15_0\rangle$	$\times$	$\times$	$\times$

Number of Infected Agents. The number of infected agents using representative vectors is calculated as:

$$\langle \hat{n} \rangle_\mu = \sum_k \langle \mu_p | \hat{n}_k | \mu_p \rangle. \tag{9}$$

In the string representation, native string methods, such as *count('x')*, count the number agents with health state $x = 0, 1, 2 \ldots$. If native methods are unavailable, one may always perform a comparative loop over the string. Algorithm A2. explains the standard procedure to count bits in the integer representation. It is worth mentioning that the operator $\sum_k \hat{n}_k$ commutes with $\hat{P}$.

Permutation. Cyclic permutations are the core transformations here. In the string representation, cyclic permutations consist of one copy and one concatenation call, as exemplified in Figure 6a.

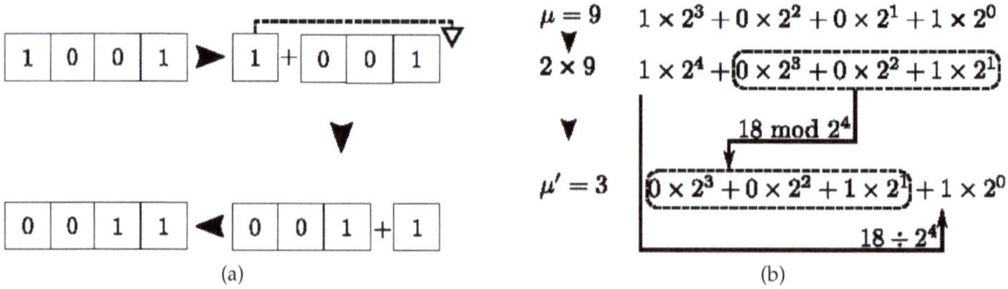

Figure 6. Cyclic permutation for configuration $\mu = 9$ with $N = 4$. (**a**) String representation executes one copy and one concatenation operation; (**b**) integer representation requires both integer division and modulo operation by 2^N.

Meanwhile, in the integer representation, cyclic permutations are obtained using modulo and integer division: $\mu' = (2\mu \,\%\, 2^N) + (2\mu // 2^N)$, the new configuration μ' is obtained from configuration μ taking the modulo of 2μ by 2^N in addition to the result of the integer division 2μ by 2^N. Multiplication by the number of available states translates bit fields to the left. The modulo operation crops contributions larger than those available to N-bit fields. Integer division $2\mu/2^N$ selects the bit associated to largest binary position and shifts it to the lowest binary position (see Figure 6b).

Next, we focus our attention on the sector with $p = 0$, which plays an important role in epidemic models (see Section 5 for further discussion). This invariant subspace holds only symmetric linear combinations of configuration vectors. Incidentally, that also means that configurations with short cycles—or large repetition numbers—can only have representative vectors with non-vanishing norms iff $p = 0$. The most important cases are: a) the all-infected configuration $|111 \cdots 1\rangle$, and b) disease-free configuration $|000 \cdots 0\rangle$. This occurs because these two configurations are invariant by every cyclic permutation available, including a single cyclic permutation (short cycle). As a direct consequence, the probability of disease eradication, $\pi_0(t)$, and the probability that the disease has infected each element of the population, $\pi_{2^N-1}(t)$, can only be evaluated at $p = 0$. Moreover, this sector holds the largest dimension being the worst scenario for numerical computations.

To construct the vectors for this particular sector, consider each integer μ in $[0, 2^N)$ as a potential candidate to assemble the symmetric vector spaces for fixed p. By performing $N-1$ cyclic permutations over $|\mu\rangle$, one determines the representative state $|\mu_p\rangle$ in Equation (5), as well as the number of repetitions $R_{\mu,p}$, hence the norm $\mathcal{N}_\mu$. Algorithm A3. calculates the representative vector $|\mu_p\rangle$ associated with configuration $|\mu\rangle$. Due to the order convention adopted here, the string representation must be converted to the integer representation at the *if*-clause test. The representative configurations are then stored either in a list or dictionary. As an additional benefit, since vector spaces are independent of the problem at hand, the set of representatives may also be stored in a database for further use in different problems, as long as they are subjected to the same symmetry.

4. Matrix Elements

The next step is the evaluation of the transition matrix in the sector $p = 0$. Infection and recovery dynamics are the main actors in this context, as they inform the way representative vectors $|\mu_0\rangle$ interact with each other, $\hat{T}|\mu_0\rangle = \sum'_{\{\nu\}} T_{\nu\mu}|\nu_0\rangle$. The prime indicates the sum runs over all eigenvectors in the $p = 0$ sector, while cyclic permutation invariance implies:

$$\hat{T}|\mu_0\rangle = \frac{1}{\mathcal{N}_\mu} \sum_{k=0}^{N-1} \hat{P}^k \, \hat{T}|\mu\rangle \, . \tag{10}$$

Equation (10) tells us the action of $\hat{T}$ on the linear combination $|\mu_0\rangle$ is calculated from the simpler operation $\hat{T}|\mu\rangle$. The resulting vectors are then permuted, producing the corresponding matrix elements. The practical advantages of this method come from the order of the operations: By doing the transitions first and then finding the respective representatives, one divides the workload by a factor N. If the normal ordering were used instead, one would evaluate the transitions for each element of the linear combination and then find the corresponding representative, hence N times the number of operations required with transition first. For instance, consider $\hat{T}|7_0\rangle$ for $N = 3$:

$$\begin{aligned}
\hat{T}|7_0\rangle =& \frac{1}{\mathcal{N}_7} \sum_{k=0}^{2} \hat{P}^k \hat{T}|7\rangle = \frac{\gamma}{\mathcal{N}_7} \sum_{k=0}^{2} \hat{P}^k \left(|3\rangle + |5\rangle + |6\rangle\right) \\
=& \frac{\gamma}{\mathcal{N}_7} \sum_{k=0}^{2} \hat{P}^k \left(|3\rangle + \hat{P}|3\rangle + \hat{P}^2|3\rangle\right) = \left(3\gamma \frac{\mathcal{N}_3}{\mathcal{N}_7}\right) |3_0\rangle \\
=& \sqrt{3}\gamma|3_0\rangle.
\end{aligned} \tag{11}$$

The relevant data structure for $\hat{T}$ are the off-diagonal transitions, which are further subdivided into two categories: one or two-body contributions. This is illustrated in Figure 7 for the SIS model. The finite set of transition rules are passed as a lookup table or, if available, a dictionary. Data is organized as follows: Each entry represents a one or two-body configuration whose value corresponds to one tuple. Each tuple holds two immutable values: the configuration to which the entry transitions to and the assigned coupling strength.

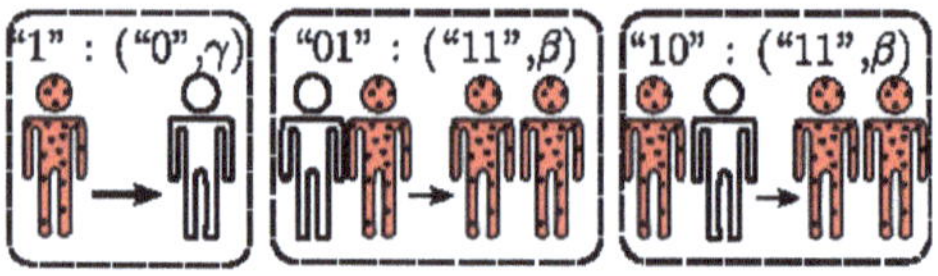

Figure 7. Off-diagonal transitions in the SIS model. Data structure follows the income-outcome convention. Data entries represent the current one-body (two-body) health state, whereas the corresponding data values, organized as tuples, express the outcome one-body (two-body) configuration and coupling strength.

With off-diagonal transition rules in hand, one-body actions are evaluated by scanning each agent and applying the corresponding transition rule in Algorithm A4.. The resulting one-body transitions are stored in the *outcome* variable. Figure 8 depicts an example for $N = 3$ and one infected agent at $k = 1$. Two-body operators differ from their one-body counterparts due to the fact they require two agent loops and information from the adjacency matrix A, as seen in Algorithm A5.. Figure 9 exhibits an example for $N = 3$. After both one- and two-body transitions are computed, the diagonal element is obtained *via* probability conservation: $T_{\mu\mu} = 1 - \sum'_{\mu \neq \nu} T_{\mu\nu}$. The process is iterated until all eigenvectors and their respective transitions are accounted.

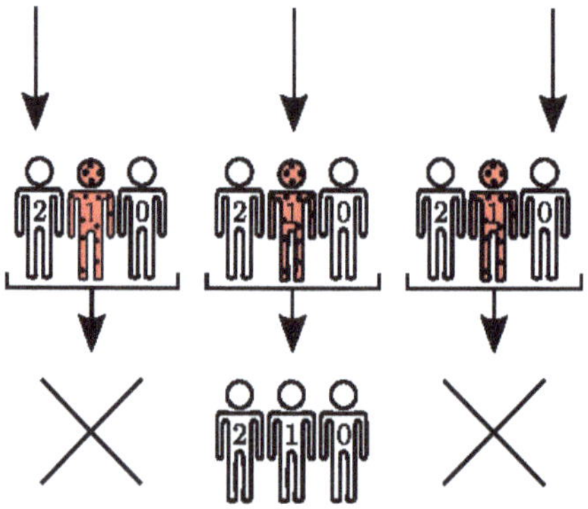

Figure 8. Recovery operator action on configuration vector $|2\rangle$, in the SIS model with $N = 3$. Non-vanishing transition is observed only for agent $k = 1$, which is infected, producing $|0\rangle$.

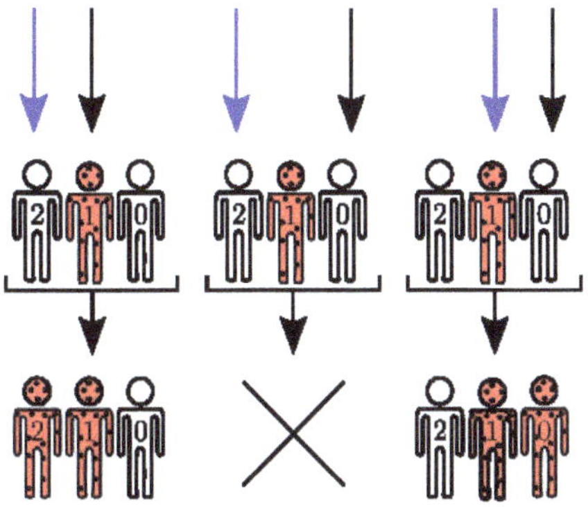

Figure 9. Infection operator action on configuration vector $|2\rangle$, in the SIS model with $N = 3$ and mean field network. Disease transmission events are evaluated for each pair of agents. Whenever the pair health state differs, and the pair also shares one connection expressed by the adjacency matrix, the configuration changes to contemplate the recently infected individual. For $|2\rangle$, $k = 1$ agent contaminates $k = 0$ ($k = 2$) agent, producing the configuration $|3\rangle$ ($|6\rangle$).

5. Casimir Vector Space

The recent advances in the disease spreading dynamics in realistic populations are intimately linked to network theory [12,24]. Networks are traditionally associated with graphs holding a large number of nodes and links [25]. The graph must be large enough to produce a degree distribution, which describes the probability distribution of links per node. The degree distribution, or alternatively its statistical moments, characterizes the network type and its properties. However, some networks, including random networks, require an ensemble of graphs to provide an accurate picture. Thus, a graph becomes a sample or realization of the network. Statistical properties of networks are derived for each graph, followed by ensemble average and deviation. In practice, when graphs in the ensemble are large enough ($N \gg 1$) and representatives, statistics may also be evaluated for each graph and extrapolated as those of the network.

Two cases hold particular importance for applications of network theory in epidemic models: the mean field and random networks. In the first case, all agents are connected, meaning one infected agent may potentially infect anyone. Hence, the disease tends to spread faster than in constrained networks. Furthermore, all graphs in the mean field ensemble share the same adjacency A^{MF}. In the other case, the connection between agent i and j occurs with fixed probability ρ. However, graphs in the random network ensemble differ from each other. Here, we only consider ensemble averages as a way to extract statistical properties, which is equivalent to set $A_{ij}^{\mathrm{random}} = \rho\,(1 - \delta_{ij}) = \rho A_{ij}^{\mathrm{MF}}$. Thus, all relevant symmetries lie only in the mean field adjacency matrix A^{MF}. Naturally, A^{MF} remains invariant under cyclic permutations, enabling the application of the algorithm explained in the previous sections. However, A^{MF} is also symmetric under the action of any permutation, which drastically reduces the diagonal blocks of $\hat{T}$ from $O(2^N/N)$ to $O(N)$.

Here, our primary concern is to employ the cyclic permutation eigenvectors $|\mu_p\rangle$ to generate the eigenvectors of the complete permutation group, $|s, m; p\rangle$. The eigenvectors $|s, m; p\rangle$ reduce $\hat{T}$ in mean field or random networks to block diagonal form with dimension $O(N)$. The indices s and m may assume the following values $s = N/2, N/2 - 1, \ldots$ with $s > 0$ and $m = -s, -s + 1, \ldots, s$, respectively. The relationship between s and m are the same as those observed for quantum spin operators. The explanation goes as follows. As shown in Reference [16], Equation (4), in either mean field or random networks, contains operators $\hat{S}^{\pm} \equiv \sum_k \hat{\sigma}_k^{\pm}$ and $\hat{n} \equiv \sum_k \hat{n}_k$. From the important relation $\hat{n} = \hat{S}^z + N/2$, one retains spin operators and the upper bound $s = N/2$, as expected from the combination of N $1/2$-spin particles.

In what follows, we only consider the $p = 0$ sector. First, let $\hat{S}^2 = (\hat{S}^z)^2 + (\hat{S}^+\hat{S}^- + \hat{S}^-\hat{S}^+)/2$ be the Casimir operator, so that $[\hat{S}^2, \hat{S}^\alpha] = [\hat{P}, \hat{S}^\alpha] = 0$ for $\alpha = z, \pm$ and $\hat{S}^2|s, m; 0\rangle = s(s+1)|s, m; 0\rangle$. Accordingly, $[\hat{S}^2, \hat{T}] = 0$ and s and p are good quantum numbers. In general, the eigenvector $|s, m; p\rangle$ may always be expressed as:

$$|s, m; p\rangle = \sum_{\mu} c_{\mu}^{smp} |\mu\rangle \ . \tag{12}$$

Clearly, $c_{\mu}^{smp} = 0$ if the number of infected agents in the configuration μ, $n_\mu = \sum_k \langle \mu_0 | \hat{n}_k | \mu_0 \rangle$, fails to satisfy the constraint $n_\mu = m + N/2$. The idea is to write Equation (12) as a linear combination of representative vectors $|\mu_p\rangle$ with $m + N/2$ infected agents, ensuring all available permutations are accounted for. The implications for numerical codes is quite obvious: it allows the reuse of numerical codes to obtain eigenvectors $|\mu_p\rangle$.

The most relevant sector for epidemic models contains the configuration with all (none) infected agents. According to previous sections, this implies $p = 0$ while $m = \pm N/2$ requires $s = N/2$. In the $(s = N/2, p = 0)$ sector, the desired linear combination is:

$$|s = N/2, m, p = 0\rangle = \frac{1}{\mathcal{N}} \sum_{\{\mu\}}' R_{\mu,0}{}^{-1/2} |\mu_0\rangle \ , \tag{13}$$

with normalization $|\mathcal{N}|^2 = \sum_{\mu}' |R_{\mu,0}|^{-1}$. The prime indicates the sum is subjected to the constraint $n_\mu = m + N/2$ for $m = -N/2, \ldots, N/2$. The result in Equation (13) agrees with the standard theory of spin addition. Generalization for p and s is straightforward and omitted. It is worth mentioning the formalism adopted here already accounted for forbidden states in $p \neq 0$ sectors.

Examples are available to appreciate Equation (13) for increasing values of N. We begin considering $N = 4$. This translates into $s = 2$ and $m = -2, \ldots, 2$. The relevant representative eigenvectors $|\mu_0\rangle$ are expressed in Table 2. The only non-trivial correspondence occurs for $m = 0$,

$$|2, 0; 0\rangle = \frac{\sqrt{2}|3_0\rangle + |5_0\rangle}{\sqrt{3}} = \frac{|0011\rangle + |1001\rangle + |1100\rangle + |0110\rangle + |0101\rangle + |1010\rangle}{\sqrt{6}} . \tag{14}$$

Next, consider $N = 6$ which fixes $s = 3$ and $m = -3, \ldots, 3$. The eigenvector $|3, 0; 0\rangle$ holds contributions from four cyclic eigenvectors or, equivalently, 20 configurations:

$$\begin{aligned} |3, 0; 0\rangle =& \frac{\sqrt{3}|7_0\rangle + \sqrt{3}|11_0\rangle + \sqrt{3}|19_0\rangle + |21_0\rangle}{\sqrt{10}} \\ =& \frac{|000111\rangle + |100011\rangle + |110001\rangle + |111000\rangle + |011100\rangle + |001110\rangle}{\sqrt{20}} . \\ +& \frac{|001011\rangle + |100101\rangle + |110010\rangle + |011001\rangle + |101100\rangle + |010110\rangle}{\sqrt{20}} . \\ +& \frac{|010011\rangle + |101001\rangle + |110100\rangle + |011010\rangle + |001101\rangle + |100110\rangle}{\sqrt{20}} . \\ +& \frac{|010101\rangle + |101010\rangle}{\sqrt{20}} . \end{aligned} \tag{15}$$

Table 2. Eigenvectors $|\mu_0\rangle$ with $N = 4$.

μ_0	$R_{\mu,0}$	m	$\|\mu\rangle$
0	4	−2	$\|0000\rangle$
1	1	−1	$\|0001\rangle$
3	1	0	$\|0011\rangle$
5	2	0	$\|0101\rangle$
7	1	1	$\|0111\rangle$
15	4	2	$\|1111\rangle$

6. Discussion

The algorithms presented in this study assumed only two health states for each agent. Generalization for q number of states is readily available by changing to the integer representation $\mu = a_{N-1}q^{N-1} + \cdots + a_0q^0$, with $a_k = 0, 1, \ldots, q-1$, concomitant with additional off-diagonal transitions. For instance, the susceptible-infected-recovered-susceptible (SIRS) ABEM generalizes the SIS model as it introduces the removed (R) health state for agents. This additional state often means the agent has recovered from the illness and developed immunity, has been vaccinated, or has passed away. In any case, once removed, the agent takes no part in the dynamics of disease transmission, hindering infection events [12]. As such, recovery with immunization or death events produce the transition $I \to R$, with probability γ while vaccination $S \to R$ occurs with probability ξ. If death events are excluded, temporary immunization is achieved via $R \to S$ with probability η. Therefore, the vectors $|n_{N-1}\cdots n_0\rangle$ with $n_k = 0, 1$ or 2 describe configurations of the SIRS model. However, the algorithm to explore cyclic permutations remains unchanged as it explores symmetries of the underlying network. As a result, eigenvalues and number of sectors are the same, but degeneracy and eigenvectors change to accommodate the increased number of health states.

Parallelism merits further discussion. The computation of representative vector space may be performed in parallel by dividing the set of q^N integers among Q processes. Each process runs one local set of representative vectors which, posteriorly, is compared against the sets from the remaining processes. The union of all Q sets produces the desired representative vector space. Parallelism is also obtained at the evaluation of $\hat{T}$: Columns ($|\mu_p\rangle$) are distributed among Q processes and the corresponding matrix elements are calculated for each process. The union of all matrix elements from each process produces the complete description of $\hat{T}$ in the representative vector space. Lastly, parallelism is also available for sparse products $\hat{T}|\pi(t)\rangle$ necessary to execute the time evolution.

We also emphasize the algorithms explained here are most useful to evaluate quantities within a single permutation sector of $\hat{T}$. This is likely the case whenever the probability for disease eradication or complete population contamination are concerned. Another relevant situation occurs when the initial condition itself falls within a single sector. For instance, the initial probability vector $|\pi(0)\rangle = (1/3)(|001\rangle + |010\rangle + |100\rangle)$ states only one among $N = 3$ agents is infected. However, the identity of the infected agent is unknown a priori, so that configurations with one infected agent occurs with equal probability $1/N$. Now, the decomposition of $|\pi(0)\rangle$ in the $|\mu_p\rangle$ basis results in $|\pi(0)\rangle = (1/\sqrt{3})|1_0\rangle$. Thus, the time evolution of $|\pi(0)\rangle$ by the action of $\hat{T}$ is again restricted to a single permutation sector.

Without loss of generality, the initial condition can always be written as $|\pi(0)\rangle = \sum'_{\{\mu\}} \sum_{k=0}^{N-1} \pi_{\mu k}\hat{P}^k|\mu\rangle$, where the primed sum runs only over the indices μ, which also labels the representative vectors. The cyclic permutation $\hat{P}^k$ generates the remaining configurations related to $|\mu\rangle$ whereas the coefficients $\pi_{\mu k}$ are the corresponding initial probabilities. Using the eigenvalue equation for $\hat{P}$, one calculates the scalar product:

$$\langle \nu_p|\pi(0)\rangle = \sum_{\{\mu\}}{}' \sum_{k=0}^{N-1} \pi_{\mu k}\langle \nu_p|\hat{P}^k|\mu\rangle = \sum_{\{\mu\}}{}' \sum_{k=0}^{N-1} \pi_{\mu k}e^{2i\pi pk/N}\langle \nu_p|\mu\rangle = \sqrt{N}\tilde{\pi}_{\nu p}\frac{R_{\nu p}}{\mathcal{N}_\mu}, \quad (16)$$

where $\tilde{\pi}_{\mu p} = N^{-1/2}\sum_k \pi_{\mu k} e^{2i\pi pk/N}$ is the discrete Fourier transform of $\pi_{\mu k}$. Using the previous example, with one infected among $N = 3$ agents, $|\pi(0)\rangle = \sum_{k=0}^{2} \pi_{0k}\hat{P}^k|0\rangle + \sum_{k=0}^{2} \pi_{1k}\hat{P}^k|1\rangle + \sum_{k=0}^{2} \pi_{3k}\hat{P}^k|3\rangle + \sum_{k=0}^{2} \pi_{7k}\hat{P}^k|7\rangle$, with $\pi_{\mu k} = \delta_{\mu 1}/3$ so that $R_{1p} = 1$, $\tilde{\pi}_{1p} = \delta_{p0}/\sqrt{3}$, and the previous result is recovered.

Now we address the case where the evaluation of the desired statistics requires several permutation sectors. In the worst case scenario, every permutation sector contributes equally to the computation. Therefore one must diagonalize each block in order to obtain the relevant eigenvalues and eigenvectors. As a crude approximation, one may consider that the N blocks have the same dimension d/N for a d-dimensional vector space. The complexity of diagonalization methods in the LAPACK library range from $O((d/N)^2)$ up to $O((d/N)^3)$ for each block [26], whereas the complexity range for full diagonalization is $[O(d^2), O(d^3)]$. Thus diagonalization of N blocks reduces the total complexity from N^{-1} up to N^{-2}. More importantly, blocks can be diagonalized in different processors because they are disjointed.

The algorithms presented here are most suitable for networks with invariance by cyclic permutations. However, they are also convenient whenever the algebraic commutator can be approximated by $[\hat{T}, \hat{P}] = \hat{O}$, where the operator $\hat{O}$ is symmetric under cyclic permutations, $[\hat{O}, \hat{P}] = 0$. In particular, $\hat{O} = q_0\mathbb{1} + q_1\hat{P}^y + \sum_{\beta=z,\pm} q_\beta \hat{S}^\beta$, with constant q_j ($j = 0, 1, z, \pm$) and $y \in \mathbb{R}$, creates interesting disease-spreading dynamics, such as a localized disease source for $q_\beta = q\delta_{\beta,0}$.

Finally, we compare performances of the SIS ABEM using the transition matrix method with and without our algorithm. Numerical experiments were performed using Python on an Intel-PC i7-7700 3.8 GHz. The decision to pick up Python instead of a more performance-oriented language was based on the ability to quickly disseminate the method. For data intensive research, we strongly recommend performance-oriented languages, such as C or high-performance Fortran. The results are summarized in Table 3. As expected, cyclic permutations greatly improve computation times, most noticeable for large populations sizes. For $N = 20$, the improved numerical code runs two orders of magnitude faster, while only consuming a fraction—about 6%—of the original memory. We reiterate methods involving the transition matrix to compute the probabilities of each configuration available to the system $\pi_\mu(t)$, with $\mu = 0, 1, \ldots, 2^N - 1$, up to numerical errors (floating point and rounding errors), often around $O(10^{-12})$. Because they include all configurations, they can provide accurate statistics and data predictions along the evolution of the epidemics. However, direct Monte Carlo methods (DMCM) are far more efficient if one is solely interested in a few statistical moments of relevant variables, not in the entire joint pdf [27,28]. There are mainly two flavors of DMCM, depending on whether the time interval is fixed or distributed according to a given PDF [29]. The latter case is more commonly known as the Gillespie algorithm [30–32], and it has been successful to simulate epidemics. In DMCMs, execution times are directly related to the number of independent runs m, with error scaling as $m^{-1/2}$. Usually, $m \sim O(10^6)$ produces errors around $O(10^{-3})$. Smaller errors can be obtained by increasing m. Regardless, DMCM are always more efficient if the joint PDF is not required, as they probe the configurations that are more likely to occur. Indeed, computation times of DMCM with $N = 20$, $\tau = 0.19$ s, are far lower than the 41 s obtained previously. Furthermore, DMCM hold small memory footprint and can simulate ABEM with $N \sim O(10^4)$.

Table 3. Computation times and memory usage of time evolution of the SIS agent-based epidemic models (ABEM) for various population sizes, with (block) and without (full matrix) cyclic permutations.

N	Time (s)		Memory (MB)	
	Block	**Full Matrix**	**Block**	**Full Matrix**
10	0.02(9)	0.45(7)	23.3(0)	24.7(1)
12	0.10(0)	2.91(4)	23.9(8)	32.3(9)
14	0.41(1)	19.59(6)	26.5(2)	62.4(8)
16	1.76(6)	132.31(5)	34.8(1)	201.0(8)
18	8.38(4)	837.86(9)	66.4(5)	784.6(8)
20	41.93(1)	4529.09(7)	184.0(4)	3251.0(7)

7. Conclusions

ABEM describe the stochastic dynamics of disease-spreading processes in networks. Direct investigation of epidemic Markov processes is often hindered due to the exponential increase of the dimension of the vector space with the number of agents. By exploiting cyclic permutation symmetries, relevant elements of the dynamics are confined to a single permutation sector, significantly reducing computational efforts. In practical terms, by selecting a single cyclic permutation eigensector, one selects only relevant information from the stochastic process. The $p = 0$ sector holds particular importance, as it contains configurations where none or all agents are infected, with dimension scaling as $O(N)$ for highly connected networks. Our findings show that using symmetric basis significantly improves computation times and reduces memory usage, providing a detailed picture of the joint probability distribution function. This development allows for a more detailed investigation of fluctuations and correlation functions in epidemics. For global statistics that describe the evolution of the epidemic, DMCM provide much faster computation times subjected to a given statistical error. In closing, the inclusion of finite symmetries brings down ABEM to the same footing of compartmental models regarding the number of agents but does not neglect the role played by fluctuations.

Author Contributions: A.S.M. and G.M.N. designed the research; A.C.P.M. and G.M.N. performed the research and wrote computational codes; A.C.P.M. verified numerical results; G.M.N. wrote the paper; G.C.C. and A.S.M. edited the paper. All authors reviewed the manuscript.

Funding: G.M.N. thanks CAPES/PNPD 88887.136416/2017-00, A.S.M. holds grants from CNPq 307948/2014-5, G.C.C. acknowledges funding from CAPES 067978/2014-01 and A.C.P.M. acknowledges grant CNPq 800585/2016-0.

Conflicts of Interest: The authors declare no conflict of interest.

Abbreviations

The following abbreviations are used in this manuscript:

ABEM	Agent-based epidemic model
NPDF	Network probability distribution function
SIS	Susceptible-infected-susceptible
SIRS	Susceptible-infected-recovered-susceptible
WHO	World health organization

Appendix A. Algorithms

Appendix A.1. Time Evolution

Algorithm A1. Time Evolution

Require: $p \in \mathbb{N}$, matrix A and off-diagonal transitions
  $S = \{\ \}$ ▷ Basis
  for $\mu = 0$ to $\mu < 2^N$ **do**
    $\psi, \mathcal{N}_\psi \leftarrow$ calculates eigenvector and norm from μ
    Add ψ to S
  end for ▷ p invariant eigensector
  for ψ in S **do**
    for $k = 0$ to $k < N$ **do**
      $\psi' \leftarrow$ off-diagonal transitions from k-th component of ψ
      Evaluate $T_{\psi'\psi}$ ▷ Sparse storage
    end for
  end for
  $\pi \leftarrow$ initial condition
  for $t = 0$ to $t < t_{\max}$ **do**
    $\pi \leftarrow \hat{T} \times \pi$
  end for ▷ End time evolution

Appendix A.2. Number of Infected Agents

Algorithm A2. Number of Infected Agents

---

function COUNT(μ,count)
  c $\leftarrow \mu$
  count $\leftarrow 0$
  for $k = 0$ to $k < N$ **do**
    count $\leftarrow$ count + c % 2
    c $\leftarrow$ c // 2
  end for
end function

Appendix A.3. Representative Vectors

Algorithm A3. Representative Vectors

---

function REPRESENTATIVE(μ, ψ, r)
  $\psi \leftarrow \mu$
  $r \leftarrow 1$
  for $k = 0$ to $k < N - 1$ **do**
    $\mu \leftarrow \hat{P}\mu$
    if $\mu < \psi$ **then**
      $\psi \leftarrow \mu$
    else if $\mu = \psi$ **then**
      $r \leftarrow r + 1$
    end if
  end for
end function

Appendix A.4. One-Body Off-Diagonal Transitions

Algorithm A4. One-Body Off-Diagonal Transitions

```
function ONEBODY(label,rules,output)
    for k = 0 to k < N do                                    ▷ Loop over agents
        if label[k] in rules then
            new ← label with label[k] ← rule[label[k]][0]
            output[new] ← coupling
        end if
    end for
end function
```

Appendix A.5. Two-Body Off-Diagonal Transitions

Algorithm A5. Two-Body Off-Diagonal Transitions

```
function TWOBODY(L,A,rules,outcome)
    for j = 0 to j < N do
        for i = 0 to i < N do
            q ← (L_j L_i)
            if q in rules then
                x ← L
                x_j ← rules[q]_00
                x_i ← rules[q]_01
                output[x] ← output[x] + A_ji
            end if
        end for
    end for
end function
```

References

1. Mlakar, J.; Korva, M.; Tul, N.; Popović, M.; Poljšak-Prijatelj, M.; Mraz, J.; Kolenc, M.; Resman Rus, K.; Vesnaver Vipotnik, T.; Fabjan Vodušek, V.; et al. Zika Virus Associated with Microcephaly. *N. Engl. J. Med.* **2016**, *374*, 951–958. [CrossRef] [PubMed]
2. Maganga, G.D.; Kapetshi, J.; Berthet, N.; Kebela Ilunga, B.; Kabange, F.; Mbala Kingebeni, P.; Mondonge, V.; Muyembe, J.J.T.; Bertherat, E.; Briand, S.; et al. Ebola Virus Disease in the Democratic Republic of Congo. *N. Engl. J. Med.* **2014**, *371*, 2083–2091. [CrossRef] [PubMed]
3. Team, W.E.R. West African Ebola Epidemic after One Year—Slowing but Not Yet under Control. *N. Engl. J. Med.* **2015**, *372*, 584–587. [CrossRef]
4. Van Kerkhove, M.D.; Bento, A.I.; Mills, H.L.; Ferguson, N.M.; Donnelly, C.A. A review of epidemiological parameters from Ebola outbreaks to inform early public health decision-making. *Sci. Data* **2015**, *2*, 150019. [CrossRef]
5. Team, W.E.R. After Ebola in West Africa — Unpredictable Risks, Preventable Epidemics. *N. Engl. J. Med.* **2016**, *375*, 587–596. [CrossRef] [PubMed]
6. Willem, L.; Verelst, F.; Bilcke, J.; Hens, N.; Beutels, P. Lessons from a decade of individual-based models for infectious disease transmission: A systematic review (2006–2015). *BMC Infect. Dis.* **2017**, *17*, 612. [CrossRef]
7. Keeling, M.; Eames, K. Networks and epidemic models. *J. R. Soc. Interface* **2005**, *2*, 295–307. [CrossRef] [PubMed]
8. Djurdjevac Conrad, N.; Helfmann, L.; Zonker, J.; Winkelmann, S.; Schütte, C. Human mobility and innovation spreading in ancient times: a stochastic agent-based simulation approach. *EPJ Data Sci.* **2018**, *7*, 24. [CrossRef]
9. Holme, P. Three faces of node importance in network epidemiology: Exact results for small graphs. *Phys. Rev. E* **2017**, *96*, 062305. [CrossRef] [PubMed]

10. de Espíndola, A.L.; Bauch, C.T.; Cabella, B.C.T.; Martinez, A.S. An agent-based computational model of the spread of tuberculosis. *J. Stat. Mech. Theor. Exp.* **2011**, *2011*, P05003. [CrossRef]
11. Gómez, S.; Arenas, A.; Borge-Holthoefer, J.; Meloni, S.; Moreno, Y. Discrete-time Markov chain approach to contact-based disease spreading in complex networks. *Eur. Phys. Lett.* **2010**, *89*, 38009. [CrossRef]
12. Pastor-Satorras, R.; Castellano, C.; Van Mieghem, P.; Vespignani, A. Epidemic processes in complex networks. *Rev. Mod. Phys.* **2015**, *87*, 925–979. [CrossRef]
13. Van Mieghem, P. The N-intertwined SIS epidemic network model. *Computing* **2011**, *93*, 147–169. [CrossRef]
14. Van Mieghem, P.; Cator, E. Epidemics in networks with nodal self-infection and the epidemic threshold. *Phys. Rev. E* **2012**, *86*, 016116. [CrossRef] [PubMed]
15. Wang, H.; Li, Q.; D'Agostino, G.; Havlin, S.; Stanley, H.E.; Van Mieghem, P. Effect of the interconnected network structure on the epidemic threshold. *Phys. Rev. E* **2013**, *88*, 022801. [CrossRef] [PubMed]
16. Nakamura, G.M.; Monteiro, A.C.P.; Cardoso, G.C.; Martinez, A.S. Efficient method for comprehensive computation of agent-level epidemic dissemination in networks. *Sci. Rep.* **2017**, *7*, 40885. [CrossRef]
17. Alcaraz, F.C.; Rittenberg, V. Directed Abelian algebras and their application to stochastic models. *Phys. Rev. E* **2008**, *78*, 041126. [CrossRef]
18. Alcaraz, F.; Droz, M.; Henkel, M.; Rittenberg, V. Reaction-Diffusion Processes, Critical Dynamics, and Quantum Chains. *Ann. Phys.* **1994**, *230*, 250–302. [CrossRef]
19. Reichl, L. *A Modern Course in Statistical Physics*; Wiley: New York, NY, USA, 1998.
20. Hamermesh, M. *Group Theory and Its Application to Physical Problems*; Courier Corporation: North Chelmsford, MA, USA, 1962.
21. Alcaraz, F.C.; Nakamura, G.M. Phase diagram and spectral properties of a new exactly integrable spin-1 quantum chain. *J. Phys. A Math. Gen.* **2010**, *43*, 155002. [CrossRef]
22. Nakamura, G.M.; Mulato, M.; Martinez, A.S. Spin gap in coupled magnetic layers. *Phys. A Stat. Mech. Appl.* **2016**, *451*, 313–319. [CrossRef]
23. Sakurai, J.J.; Tuan, S.F. *Modern Quantum Mechanics*; Addison-Wesley: Boston, MA, USA, 1994.
24. Bianconi, G. Interdisciplinary and physics challenges of network theory. *Eur. Phys. Lett.* **2015**, *111*, 56001. [CrossRef]
25. Newman, M.E.J. The Structure and Function of Complex Networks. *SIAM Rev.* **2003**, *45*, 167–256. [CrossRef]
26. Demmel, J.W.; Marques, O.A.; Parlett, B.N.; Vömel, C. Performance and accuracy of LAPACK's symmetric tridiagonal eigensolvers. *SIAM J. Sci. Comput.* **2008**, *30*, 1508–1526. [CrossRef]
27. Nakamura, G.; Gomes, N.; Cardoso, G.; Martinez, A. Robust Parameter Determination in Epidemic Models with Analytical Descriptions of Uncertainties. *arXiv* **2018**, arXiv:1807.05301.
28. Nakamura, G.M.; Gomes, N.D.; Cardoso, G.C.; Martinez, A.S. Numerical data and codes for: Improved SIS epidemic equations based on uncertainties and autocorrelation functions. *OSF Digital Repository* **2019**. [CrossRef]
29. Thijssen, J. *Computational Physics*; Cambridge University Press: Cambridge, UK, 2007.
30. Gillespie, D.T. A general method for numerically simulating the stochastic time evolution of coupled chemical reactions. *J. Comput. Phys.* **1976**, *22*, 403–434. [CrossRef]
31. Fennell, P.G.; Melnik, S.; Gleeson, J.P. Limitations of discrete-time approaches to continuous-time contagion dynamics. *Phys. Rev. E* **2016**, *94*, 052125. [CrossRef] [PubMed]
32. Vestergaard, C.L.; Génois, M. Temporal Gillespie Algorithm: Fast Simulation of Contagion Processes on Time-Varying Networks. *PLOS Comput. Biol.* **2015**, *11*, 1–28. [CrossRef] [PubMed]

Mathematical and Computational Applications

MDPI

Article

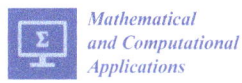

Influence of a Modulated Parameter on Hantavirus Infection

María Susana Torre [1], Anais Acquaviva [2], Jean-Marc Boyer [2] and Jorge Tredicce [2,3,*]

[1] Instituto de Física Arroyo Seco (IFAS), Universidad Nacional del Centro de la Provincia de Buenos Aires (UNCPBA), Tandil, Argentina

[2] Institut de Sciences Exactes et Appliquées (ISEA), Université de la Nouvelle Calédonie, 98851 Nouméa, Nouvelle Calédonie, France

[3] Departamento de Física, Facultad de Ciencias Exactas y Naturales, Universidad de Buenos Aires, Intendente Guiraldes 2160, Ciudad Autónoma de Buenos Aires, Argentina

* Correspondence: jorge.tredicce@inphyni.cnrs.fr

Received: 31 May 2019; Accepted: 8 July 2019; Published: 10 July 2019

Abstract: We study the dynamical behavior of a model commonly used to describe the infection of mice due to hantavirus (and, therefore, its possibility of propagation into human populations) when a parameter is changed in time. In particular, we study the situation when the ecological conditions (e.g., climate benignity, food availability, and so on) change periodically in time. We show that the density of infected mice increases abruptly as the parameter crosses a critical value. We correlate such a situation with the observed sudden outbreaks of hantavirus. Finally, we discuss the possibility of preventing a hantavirus epidemic.

Keywords: nonlinear dynamics; delay bifurcation; population dynamics

PACS: 05.45; $-a$; 64.60.Ht; 87.23.n

1. Introduction

The mouse colilargo (*Oligoryzomis longicaudatus*), unlike the other group of rodents present in South America since the arrival of Europeans, does not hibernate. Thus, it devotes all its leisure time to the intensive proliferation of its species. The colilargo is indicated as a carrier of the Hanta "Andes" virus, the strain of hantavirus that is found in the Andean–Patagonian zone. Numerous observations [1–3] carried out in several areas have shown a marked increase in the population of this rodent when there are positive ecological conditions (e.g., climate benignity and increased food availability). In this case, population explosions of these so-called "rats" may occur. The population of rodents increases notoriously because these sub-species respond quickly to the supply of food and to benevolent climate conditions. In some cases, the population density of colilargo (with normal values of 10 to 100 individuals per hectare) can rise to about 1000 to 1500 individuals per hectare [4,5]. This overpopulation generates stress among the mice, due to agglomeration and competition for food, which makes them more aggressive, generating many fights with bite wounds, which increases the propagation of the hantavirus. This behavior has been cited as the reason explaining the significant correlation between the index of abundance and the number of positive animals detected. The proportion of seropositivity can increase from 5% up to 10% of the total population of the mice [5]. High population densities also lead them to propagate outwards in space, looking for food or just better living conditions. Different types of rodents transmit the hantavirus in the different geographic areas of the world, and all must be considered potentially dangerous. Hantavirus, although with a low probability of infection, is not unimpressive in terms of its mortality. The mortality rate for humans is around 40% and it has not been possible to obtain a successful vaccine for this disease thus far [5]. Therefore, a good knowledge on the dynamics of the colilargo population could help to predict changes in the risks of human hantavirus infection and generate prevention policies. A mathematical model has already been proposed to analyze the propagation of hantavirus [6,7]. The model was based on the population dynamics of the mice and studied the evolution of the populations of healthy and infected mice. in [6], a study of spatial effects through the diffusion of mice (i.e., diffusion mainly characterized their movement through space) was carried out; additionally, a random variation of a parameter was included. In [8], the authors showed that the inclusion of the movement of the mice with respect to space (i.e., the diffusion term) affected additional features of the simulation in a physically understandable manner, with higher diffusion constants leading to greater agreement with the mean field results. Here, we show, instead, that diffusion is not necessary in order to explain a sudden increase in the density of infected mice and, as a consequence, the appearance of an epidemic of hantavirus. As the influence of the environmental conditions play a role in the evolution of the population, both at the seasonal level and at the level of very long cycles, we propose a model taking into account a variable parameter to analyze the population dynamics of the colilargo mice. We analyze the dynamical solutions and we show how the number of infected mice increases abruptly when a threshold of the control parameter is crossed. Such an increase may generate a large expansion in the transmission of the disease, even without the existence of diffusion.

2. Model and Results

If we take into account only the temporal evolution of the susceptible M_s and infected M_i mice, the corresponding differential equations (as introduced in [6,7]) are:

$$\begin{cases} \frac{dM_s}{dt} = (b-c)M_s + bM_i - (a+\frac{1}{K})M_sM_i - \frac{1}{K}M_s^2 \\ \frac{dM_i}{dt} = -cM_i + (a-\frac{1}{K})M_sM_i - \frac{1}{K}M_i^2 \end{cases}, \tag{1}$$

where the parameters b and c are the natural rates of birth and death of the susceptible and infected mice, respectively; a is the infection rate of the susceptible mice that become infected due to an encounter with an infected mouse; and the parameter K, in both equations, takes into account the limitations in the process of population growth due to competition for shared resources, which called the carrying capacity and is defined, for each biological species, as the maximum population size of the species that the environment can sustain indefinitely, in accordance with their necessities. It is well-known that the infection is chronic: Infected mice do not die of it, and they do not lose their infectiousness (probably for their whole life). Therefore, the rate c is the same for both categories of mice. It is worthwhile to note that all mice are born susceptible at a rate proportional to the total number of mice, since all mice contribute equally to procreation. Even if these equations allow for a simple interpretation of each term, it is convenient to write them in terms of the total number of mice $M = M_s + M_i$ and the infected mice M_i. By just adding the two differential equations above and replacing M_s in terms of M and M_i, we get:

$$\begin{cases} \frac{dM}{dt} = (b-c)M - \frac{1}{K}M^2 = F(M, M_i) \\ \frac{dM_i}{dt} = -cM_i + (a-\frac{1}{K})MM_i - aM_i^2 = G(M, M_i) \end{cases}. \tag{2}$$

If the variable M is independent of M_i, then it is trivial to find the stationary solutions and their stability.

The total number of mice takes only two steady-state values:

$$M = 0$$
$$\text{and} \quad M = K(b - c).$$

The first one indicates that such a type of mouse does not exist in the region and the second one is proportional to the difference between birth and death rates, where the constant of proportionality is the carrying capacity. In a phase space portrait, the two straight lines corresponding to the possible values of M are two of the nullclines (zero-growth isoclines) of the dynamical system. The equation corresponding to the infected mice immediately gives the following nullclines:

$$M_i = 0, \quad \text{and}$$
$$M_i = (a - \frac{1}{K})M - c.$$

The four nullclines are represented in Figure 1 in the space of M_i as a function of M. Their intersections are the fixed points of the system:

$$(M, M_i)_1 = (0, 0)$$
$$(M, M_i)_2 = (0, \frac{-c}{a})$$
$$(M, M_i)_3 = (K(b - c), 0)$$
$$(M, M_i)_4 = (K(b - c), K(b - c) - \frac{b}{a})$$

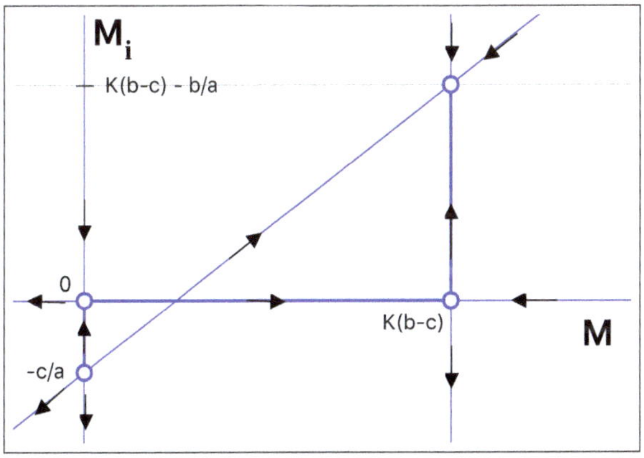

Figure 1. Nullclines in phase space. Their intersection defines the four fixed points of the system. The arrows indicate the stable and unstable manifold corresponding to each fixed point. The parameters were chosen such that the fixed point with a positive number of infected mice is stable. The fixed points $(0, 0)$ and $[K(b - c), 0]$ are both saddle points.

The first steady-state solution is the trivial one, in which there are no mice in the region of interest. The second steady-state solution, $(M, M_i)_2 = (0, \frac{-c}{a})$, is not compatible with the problem as the number of infected mice is negative. The third and the fourth solutions are interesting solutions. The third solution corresponds to a situation in which all mice are healthy, and the number of infected mice vanishes. For the solution $(M, M_i)_4 = (K(b-c), K(b-c) - \frac{b}{a})$ susceptible and infected mice co-exist, with $M_s = \frac{b}{a}$.

A very simple linear stability analysis gives the region in parameter space where one of the solutions will prevail. The Jacobian Γ of the system is:

$$\Gamma = \begin{bmatrix} \partial_M F(M, M_i) & \partial_{M_i} F(M, M_i) \\ \partial_M G(M, M_i) & \partial_{M_i} G(M, M_i) \end{bmatrix} = \begin{bmatrix} (b-c) - 2\frac{1}{K}M & 0 \\ (a - \frac{1}{K})M & -c + (a - \frac{1}{K})M - 2aM_i \end{bmatrix}.$$

Then, the eigenvalues of Γ corresponding to the $(0,0)$ solution are:

$$\lambda_1 = b - c$$
$$\text{and} \quad \lambda_2 = -c$$

Thus, as expected, the trivial solution is stable if the death rate is bigger than the birth rate, which causes the extinction of the mice. If the birth rate is greater than the death rate, the solution is clearly a saddle point. The eigenvalues for the $(K(b-c), 0)$ solution are:

$$\lambda_1 = -(b - c)$$
$$\text{and} \quad \lambda_2 = Ka(b - c) - b$$

This solution will be stable if the birth rate b remains in the range:

$$c < b < c/(1 - (1/Ka)).$$

For values of the Ka product greater than one, there exists a range of the birth rates for which the solution with all mice healthy is stable. It is important to notice that the upper limit depends on the value of K, and that the range of stability is reduced for large values of K. Outside the range of stability, the solution becomes a saddle point. Finally, the stationary solution $(K(b-c), K(b-c) - \frac{b}{a})$, which predicts the co-existence of healthy and infected mice, is stable if $b > c/(1 - (1/Ka))$. At a fixed value of the birth rate $b > c$, there exists a critical value of K at which there is an exchange of stability between the last two steady-state solutions. At that point, we have a transcritical bifurcation, leading to the appearance of infected mice. The critical value of the carrying capacity K_c is given by:

$$K_c = b/[a(b - c)].$$

We show the two relevant stationary solutions as a function of K for a birth rate bigger than the death rate in Figure 2. The bifurcation happens at the exact point where the number of infected mice becomes positive. Above the bifurcation point, the number of infected mice is proportional to the carrying capacity. The transcritical bifurcation is a smooth transformation and, therefore, it predicts a relatively slow increase in the number of infected mice as the carrying capacity K is swept across the critical value. As stated above, the mouse colilargo does not hibernate; thus, we can consider the birth rate, as well as the death rate, to be almost constant in time. We will assume, also, that the the rate of infection per mouse (a) is constant. If the capacity K is suddenly increased from an initial value K_{init} below the threshold to a final value K_{fin} above the threshold, the total number of mice increases relatively fast, from the stationary solution corresponding to K_{init} towards the steady-state solution corresponding to the final value of K. The number of infected mice will increase from 0 to the steady-state value with a well-marked lethargy, as can be seen in Figure 3. The bifurcation delay is a well-known effect each time a parameter is swept across a bifurcation point [9,10], which is a consequence of the critical slowing down at the bifurcation point [11]. The delay time measured when the parameter changes discontinuously corresponds to the minimum delay. As long as the speed at which the parameter changes decreases, the delay time increases.

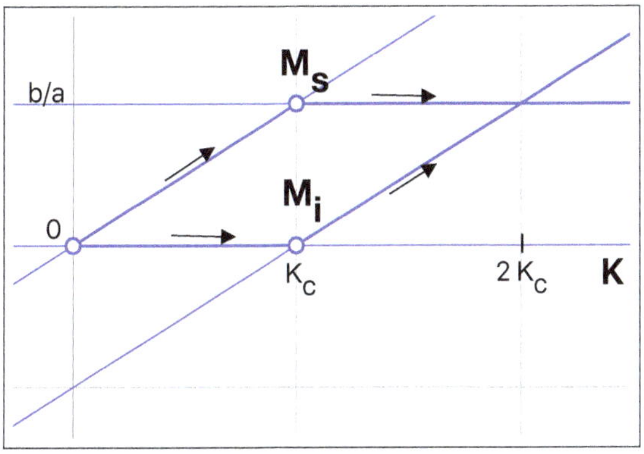

Figure 2. Steady-state values of susceptible mice (M_s) and infected mice (M_i) as a function of the carrying capacity K. Note the transcritical bifurcation at $K = K_c$. The infected mice stationary solution switches from 0 to a linear increase with K, while the number of susceptible mice switches from a linear growth with K to become constant. Therefore, after the bifurcation, the increase in the carrying capacity has, as a consequence, an increase of infected mice. The arrows in the figure show the evolution of both populations as K is increased.

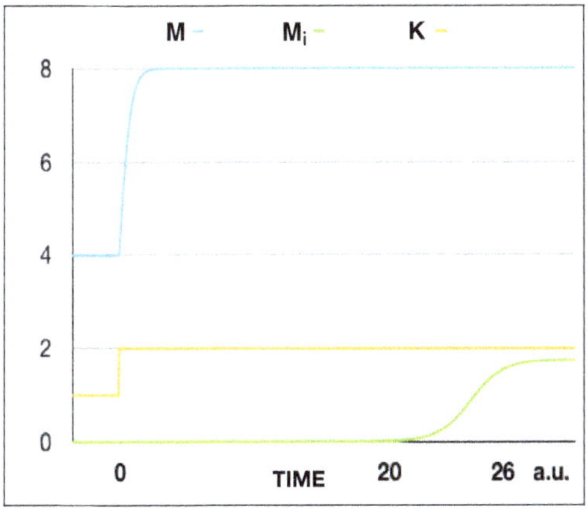

Figure 3. The total number of mice (M), infected mice (M_i), and carrying capacity K as a function of time t for $a = 0.8$, $b = 5$, and $c = 1$. The carrying capacity changes discontinuosly from $K = 1$ (below the threshold) to $K = 2$ (above the threshold) at $t = 0$. The total number of mice grows from $M = 4$ to $M = 8$ in a short time after $t = 0$. The number of infected mice, instead, begins to grow from a vanishing value after a time $t = 20$ and, then, grows until the stationary value $M_i = 1.75$. This graph presents evidence for a delay in the bifurcation of M_i. In fact, M_i begins to grow much later than the time at which M and K arrive at their steady-state values.

Here, we analyze the behavior of the dynamical system when the the carrying capacity K changes relatively slowly, compared to the characteristic time of the variables. We introduce the following temporal variation of the carrying capacity:

$$K = K_0(1 + m\cos(\omega t)). \tag{3}$$

Even if a periodic and smooth modulation of the carrying capacity does not strictly adhere to reality, it allows us to understand the origins of the different possible dynamical behaviors and, therefore, to correlate them to the real observations. In order to make the simulation realistic, we would need to introduce a noise term on the temporal evolution of the carrying capacity, because it is affected by several external factors which change from season to season. However, it is not the objective of this manuscript to compare numerical results with quantitative data, but, instead, to understand the qualitative processes in the dynamics of mouse populations. We analyze three different situations corresponding to different values of K_0 and m. In the first one, K_0 is smaller than K_c and m is such that the maximum value of K is still smaller than K_c. In the second one, K_0 is greater than K_c and the minimum value of K is still larger than K_c. Finally, the third is a situation in which K is swept across the critical value. In the first case, the number of healthy mice is modulated while the number of infected mice vanishes independently of the initial condition. During the whole modulation period, the solution with no infected mice remains stable. In this case, a seed of infected mice will vanish independently of the value of K. In the second case, there is always a positive number of infected mice, which becomes modulated as well as the total population of mice. The modulation follows the modulation of the carrying capacity with a different phase. The last case is the most interesting one, because the carrying capacity is swept across the bifurcation point and the system has to switch between the two solutions (which are alternating their stabilities). In this paper, we analyze the behavior of the number of mice when the frequency of the modulated parameter is smaller than the response rate of the variables. It is worthwhile to note that the frequency of the modulation, together with the amplitude of the modulation, will define the average speed at which the parameter is changed; thus, they will determine the delay time in the bifurcation. The most noticeable result consists of the fact that the number of infected mice will not increase continuously from 0 to a value that will follow the modulation. In fact, by simple observation of Figure 4, it is clear that, at a certain value of K, the number of infected mice will grow discontinuously. This behavior is more appropriate of a saddle-type bifurcation than a transcritical bifurcation. As a consequence, the number of susceptible mice M_s decreases abruptly while, at the same time, the number of infected mice M_i increases. It is evident that a graph of M_i as a function of K will show a bistable behavior as K swept across the bifurcation up and down (Figure 5). It is important to remark that the bistable behavior is a dynamical one. If the sweeping is stopped at any moment, the system will evolve towards the corresponding stable steady-state solution.

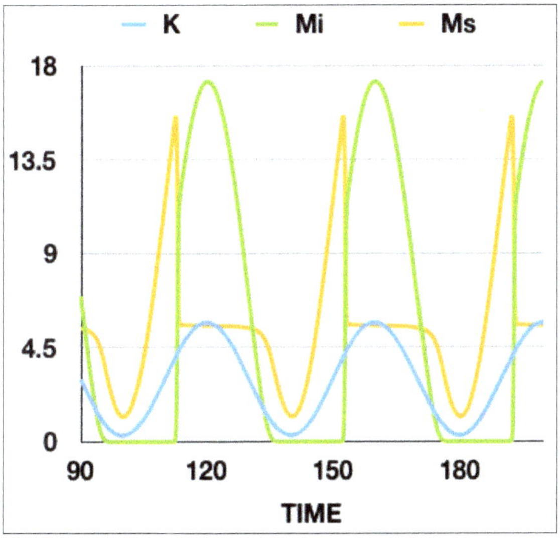

Figure 4. Number of infected mice (M_i), susceptible mice (M_s), and carrying capacity K as a function of time for $a = 0.8$, $b = 5$, $c = 1$, $K_0 = 3$, $m = 0.8$, and $\omega = 0.1571$. The behavior of the system is periodic in time. This graph presents evidence of a discontinuous increase in the number of infected mice and decrease in the number of susceptible mice when K is increased.

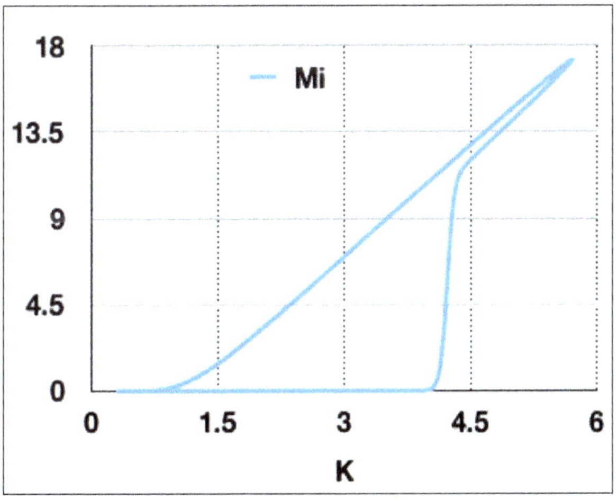

Figure 5. Infected mice (M_i) as a function of carrying capacity K corresponding to the parameter values used for Figure 4. The number of infected mice grows very fast as K increases, while it decreases slowly as K decreases. This graph presents evidence for a delay in the bifurcation of M_i. In fact, M_i begins to grow when K has already overcome the critical value corresponding to the transcritical bifurcation.

It is important to notice that this type of dynamical behavior is general and is a simple consequence of the critical slowing down at the bifurcation point. It happens by modulating every parameter and appears in a very large range of frequencies and amplitudes of the modulation, because it is intrinsically associated to the existence of a bifurcation point. The behavior will be different only if the modulation period is smaller than the decay time of the variable. A simple analysis of Figure 4 allows us to conclude that, in this simple model, a sudden increase in the infected mice is a consequence of the sweeping of a control parameter across a bifurcation point. Thus, the variable M_i does not adiabatically follow the change of the parameter, even if the rate of change of the parameter is slower than the response rate of the variable. It is worthwhile to notice that, once the number of infected mice switches on, the number of susceptible mice remains constant and the dynamical system adiabatically follows the increase of the capacity by increasing the number of infected mice. During the time that K is decreasing, the system follows almost adiabatically the evolution of the parameter, with a very small delay. The consequence of this delay is that the number of infected mice vanishes at values of K slightly smaller than the one corresponding to the bifurcation point.

3. Discussion

The results shown above suggest that the capacity K may control the appearance and disappearance of hantavirus infection in rats. In fact, if K is always below K_c, only susceptible mice exist. If just a few mice get infected, this perturbation will decay faster towards the situation where infected mice vanish. If K is always above the critical value, then infected mice will always exist. If the capacity oscillates around the Kc value, then the variables M_i and M_s do not adiabatically follow the change in K. The delayed bifurcation generates an almost discontinuous increase of the infected mice and, therefore, this effect can be understood as the origin of an outbreak of hantavirus infection among the mice with the consequence of an epidemic among other living mammals and, in particular, humans beings. On the other hand, the delay in the bifurcation justifies the infection disappearing for long periods of time, even if the carrying capacity is already above the critical value K_c; however, a further increase in K will trigger a almost immediate epidemic in the population of mice. Furthermore, it is necessary to reduce K below the critical value K_c in order to be able to eradicate the disease again. Finally, we re-affirm the importance of the non-intuitive concept that slow variations of some ecological parameters can trigger an outbreak of hantavirus infection. This concept should be taken into account in the development of public health policies.

4. Conclusions

In conclusion, we propose that a typical ecological model representing the temporal evolution of mouse population densities can explain several aspects of the propagation of hantavirus. A relatively fast variation of the parameters controlling the carrier capacity K, or of any of the parameters of the system, will produce continuous variations in their densities and a continuous, but predictable, increase in infection. However, a changing parameter which varies much slower may cause a much faster increase in the density of infected mice, generating an epidemic of hantavirus in mice. The ideal situation is represented by the situation in which the environmental conditions maintain the system below the threshold of the transcritical bifurcation. In this situation, the density of infected mice can not grow and it will always decay to a vanishing quantity. In this paper, we studied the effects of a slow variation of the environmental conditions for the mice. As a future work, it will be necessary to take into account the situations in which the living conditions change very quickly, compared to the response time of the system, and also the transmission of hantavirus among human beings.

Author Contributions: All authors contribute equally in the conceptualization, methodology and validation of the results; formal analysis, M.S.T. and J.-M.B.; numerical work, M.S.T., A.A., and J.T.

Funding: This research received no external funding.

Acknowledgments: Jorge Tredicce acknowledges discussions with Gabriel Mindlin of the University of Buenos Aires.

Conflicts of Interest: The authors declare no conflict of interest.

References

1. Schmaljohn, C.; Hjelle, B. Hantaviruses: A global disease problem. *Emerg. Infect. Dis.* **1997**, *3*, 95. [CrossRef] [PubMed]
2. Mills, J.N.; Yates, T.L.; Ksiazek, T.G.; Peters, C.J.; Childs, J.E. Long-term studies of hantavirus reservoir populations in the southwestern United States: Rationale, potential, and methods. *Emerg. Infect. Dis.* **1999**, *5*, 95. [CrossRef] [PubMed]
3. Mills, J.N.; Ksiazek, T.G.; Peters, C.J.; Childs, J.E. Long-term studies of hantavirus reservoir populations in the southwestern United States: A synthesis. *Emerg. Infect. Dis.* **1999**, *5*, 135. [CrossRef] [PubMed]
4. Philippi, F. A plague of rats. *Nature* **1879**, *20*, 530. [CrossRef]
5. Jaksic, F.M.; Lima, M. Myths and facts on ratadas: Bamboo blooms, rainfall peaks and rodent outbreaks in South America. *Austral Ecol.* **2003**, *28*, 237. [CrossRef]
6. Abramson, G.; Kenkre, V.M. Spatiotemporal patterns in the hantavirus infection. *Phys. Rev. E* **2002**, *66*, 011912. [CrossRef] [PubMed]
7. Buceta, J.; Escudero, C.; de la Rubia, F.J.; Lindenberg, K. Outbreaks of Hantavirus induced by seasonality. *Phys. Rev. E* **2004**, *69*, 021906. [CrossRef]
8. Aguirre, M.A.; Abramson, G.; Bishop, A.R.; Kenkre, V.M. Simulations in the mathematical modeling of the spread of the Hantavirus. *Phys. Rev. E* **2002**, *6*, 041908. [CrossRef]
9. Mandel, P.; Erneux, T. Laser Lorenz equations with a time-dependent parameter. *Phys. Rev. Lett.* **1984**, *53*, 1818. [CrossRef]
10. Scharpf, W.; Squicciarini, M.; Bromley, D.; Green, C.; Tredicce, J.R.; Narducci, L.M. Experimental observation of a delayed bifurcation at the threshold of an argon laser. *Opt. Commun.* **1987**, *63*, 344–348. [CrossRef]
11. Tredicce, J.R.; Lippi, G.L.; Mandel, P.; Charasse, B.; Chevalier, A.; Picqué, B. Critical slowing down at a bifurcation. *Am. J. Phys.* **2004**, *72*, 799–809. [CrossRef]

MDPI
St. Alban-Anlage 66
4052 Basel
Switzerland
Tel. +41 61 683 77 34
Fax +41 61 302 89 18
www.mdpi.com

Mathematical and Computational Applications Editorial Office
E-mail: mca@mdpi.com
www.mdpi.com/journal/mca

www.ingramcontent.com/pod-product-compliance
Lightning Source LLC
LaVergne TN
LVHW071956080526
838202LV00064B/6767
* 9 7 8 3 0 3 9 2 1 5 0 3 4 *